CNC Machining & Turning Center Programming and Operation: Including Quality in Manufacturing

Kelly Curran

Jon Stenerson

ISBN-13: 978-1533657893
ISBN-10: 1533657890

DEDICATION

To my first grandchild Nora, I love you.
Kelly Curran

To my wonderful wife, Jane
Jon Stenerson

CONTENTS

ACKNOWLEDGMENTS

We would like to acknowledge Haas Automation Inc. for their assistance with this book.

Chapter 1

INTRODUCTION TO COMPUTER NUMERICAL CONTROL

INTRODUCTION

American manufacturers have worked diligently to increase productivity, quality, process capability, reliability, and flexibility. In order to be competitive with foreign manufacturers American manufacturers have had to improve quality and efficiency. This meant using technologies that could improve quality and productivity.

OBJECTIVES

Upon completion of this chapter, the reader will be able to:

- Identify the different types of numerical control machines and their parts.
- Describe the difference between point-to-point control and continuous path control.
- Name three ways to load a program into a machine control.
- Describe the different axis coordinate systems.
- Identify positions on a Cartesian coordinate grid using absolute and incremental programming methods.

History of Computer Numerical Control (CNC)

Evolution of CNC Machines

Numerical control is nothing new. As early as 1808, weaving machines used metal cards with holes punched in them to control the pattern of the cloth being produced. Each needle on the machine was controlled by the presence or absence of a hole on the punched cards. The cards were the program for the machine. If the cards were changed, the pattern changed.

The player piano is also an example of numerical control. The player piano used a roll of paper with holes punched in it. The presence or absence of a hole determined if that note was played. Air was used to sense whether a hole was present.

The invention of the computer was one of the turning points in numerical control. In 1943 the first computer, called ENIAC (Electronic Numerical Integrator and Computer), was built.

The ENIAC computer was very large. It occupied more than 1,500 square feet and used approximately 18,000 vacuum tubes to do its calculations. The heat generated by the vacuum tubes was a constant problem. The computer could operate only a few minutes without a tube failing. In addition, the computer weighed many tons and was very difficult to program.

ENIAC was programmed through the use of thousands of switches. A $10 calculator is much more powerful than this early computer. The real turning point in computer technology was the invention of the transistor in 1948. The transistor was the replacement for the vacuum tube. It was very small, cheap, dependable, used very little power, and generated very little heat: the perfect replacement for the vacuum tube. The transistor did not see much industrial use until the 1960s.

Development of the Computer Controlled Machine

Until the 1950's complex aircraft parts were made by manual machining methods and inspected by comparing the parts to templates. The templates also had to be manufactured by manual methods, which was very time consuming and inaccurate.

In Traverse City, Michigan, a man named John Parsons was working on a method to improve the production of inspection templates for helicopter rotor blades. Parsons started as a tool room apprentice and had no college degree. Parsons' method involved calculating the coordinate points along the airfoil surface. By calculating a large number of intermediate points and then manually moving the machine tool to each of these points, the accuracy of the templates was improved. Parsons came up with the idea of using punched cards for the calculations. The data could then be used to position the machine tool. Parsons submitted a proposal to the Air Force to develop a machine to produce these templates and received a development contract in 1948. His first attempts at automatic position control used punch card tabulating machines to calculate the positions along the airfoil curve and a manual milling machine to position the tool to the calculated positions. He had two operators, one to move each axis of the machine. This method produced airfoils tens of times more accurate than the preceding method but was still a very time-consuming process.

In 1949, the Air Force awarded Parsons a contract to produce a control system that could move the axis of a machine to the calculated points automatically. The Massachusetts Institute of Technology (MIT) was subcontracted by Parsons to develop a motor that could control the axis of the machines. The servo motor was born.

Parsons envisioned a system that would calculate the path that the tool should follow and store that information on punched cards. A reader at the machine would then read the cards. The machine control would take the data from the reader and control the motors attached to each axis.

In 1951, MIT was awarded the prime contract to develop the machine control. The first machine produced by Parsons and MIT was demonstrated in 1952. Called a Cincinnati Hydrotel, it was a three-axis vertical spindle milling machine. The machine control used vacuum tubes.

One of the first attempts at making programming easier for people was called APT (Automatically Programmed Tool) Symbolic Language.

APT, invented in 1954, used English-like language to produce a program that the machine tool could understand. Remember, a machine needs the geometry of the part and machining instructions such as speeds, feeds, and coolant to operate. APT made it easier for people to write these programs, which were then translated to a program that the machine could understand.

In 1955, the Air Force awarded $35 million in contracts to manufacture numerical control machines. The first numerically controlled machine tools were very bulky. The machine control was vacuum tube–operated and needed a separate computer to generate its binary tape codes. (Binary coding systems use 1s and 0s.) Programming complex parts took highly specialized people. Developments and refinements continued, and by the early 1960s numerical control machines became much more common in industry. As the acceptance of numerical control machines grew, they became easier to use and more powerful.

In 1959, a new technology emerged: integrated circuits (ICs). Integrated circuits were actually control circuits on a chip. When manufacturers discovered how to miniaturize circuits, it helped reduce the size and improve the dependability of electronic control even more than the transistor had. Large-scale integrated circuits first were produced in 1965.

In 1974, the microprocessor was invented. This made the microcomputer, and thus small applications, possible. Great strides in the manufacture of memory for computers helped make computers more powerful and affordable.

Up until about 1976, these computer controlled machines were called NC (numerical control) machines. In 1976, CNC (computer numerical control) machines were produced. These machine controls used microprocessors to give them additional capability. They also featured additional memory. The NCs read one program step (block) at a time and executed it. CNC machines store and execute whole programs.

Types of CNC MachinesThe most common of these machine tools are machining centers (milling machines) and turning centers (lathes). These two types represent more than half of the CNC machines on the market. Other types of CNC machine tools computer include: electrical discharge machines (EDMs); CNC grinders; CNC saws; water-jet cutting machines; and coordinate measuring machines. The basic programming is very similar for all types of CNC machines. The standards and codes developed in the early years of numerical control still apply to CNC machines today.

Vertical Machining Centers
Vertical machining centers are vertical milling machines that use CNC controls for positioning and automatic tool changing to produce complex machine parts (see Figure 1–1).

FIGURE 1–1. CNC vertical machining centers use automatic tool changing systems to complete multiple machining processes.

Horizontal Machining Centers
Horizontal machining centers are horizontal milling machines that are CNC controlled (see Figure 1–2). Horizontal machining centers are equipped with automatic tool changers along with a variety of other features to increase their versatility and production capabilities. Horizontal machines are often larger and more rigid enabling heavier, faster machining.

FIGURE 1–2. A CNC horizontal machining center. Courtesy Haas Automation Inc.

CNC Turning Center

CNC turning centers are lathe-type machines and have between two and four axes (see Figure 1-3). The standard configuration consists of a two-axis lathe with one axis parallel to the spindle (Z) and one axis perpendicular to the spindle (X). Both axes are fully controllable to allow turning of chamfers and radii with standard tooling. A four-axis turning center has a rear or second tool turret that can be programmed independently of the master tool turret.

FIGURE 1–3. CNC Turning center with a bar feeder. Courtesy Haas Automation Inc.

Mill-Turn Centers

With the development of the mill-turn, the two-axis lathe becomes a three-axis machine, with the third axis being the rotation of the spindle. Besides controlling spindle speed, the programmer can control the position of the spindle radially, similar to a rotary table. When combined with a powered milling or drilling attachment (live tooling), the three-axis turning machine can now do secondary operations such as slotting and off-center drilling.

CNC Laser Cutting Machines

Laser cutting machines use coherent light as a cutting tool (see Figure 1–4). Laser cutting machines can rapidly cut plate stock into very accurate and intricate shapes.

FIGURE 1–4. Computer numerical control laser cutting machines couple the speed and accuracy of numerical control with the versatility of laser technology.

Waterjet Machines

Waterjet machines use a very high pressure stream of water and abrasive as the cutting tool (see Figure 1–5). Waterjet machines can cut plate stock into accurate intricate shapes.

FIGURE 1–5. Waterjet cutting machines couple the speed and accuracy of numerical control with the versatility of laser technology.

CNC Press Brakes

Computer-controlled press brakes are used to bend and form sheet metal into parts. For example, the steel control cabinet for a CNC machine is probably formed on a CNC press brake. These machines are programmed to control the down stroke of the ram, which helps control the angle of the bend. Interchangeable dies on the bottom also control the angle, and programmable back stops control how the work is positioned before the bend. They are programmed in step-like fashion. The operator puts the sheet metal into the press and makes a bend. The operator then repositions the work, and another bend is made. The control prompts each step. In the case of a control cabinet, the seams are then welded together, perhaps by a robot welder.

CNC Punch Press

In the past, a die had to be made for stamped parts. This die was good for only one purpose and required a long lead time. Dies were exceptionally costly for large parts. If changes were made, the die had to be reworked at great expense of money and time.

CNC punch presses are programmable. Many punches of different shapes and sizes are built into the machine. A CNC program controls the machine, which actually moves the sheet metal to the punches. The punches are used to punch holes, slots, or other internal part shapes or features and to cut the outside shape of the part, small or large. The only lead time required is to create the program. If a change is desired, the program is changed. This has drastically changed the metal-stamping business. Programming software can optimize the job so the maximum number of parts can be made from the piece of material with minimum waste, which is called *part nesting*.

Point-To-Point vs. Continuous Path

Many of the early machines were point-to-point machines.

In some machining operations, we do not care about the path; we only care about the destination. For example, consider the drilling of holes. The important thing is that the holes end up in the correct location. We don't care how the machine got from the first hole to the second, and so on. Many of the early numerical control drilling machines were point-to-point machines.

On the other hand, imagine that you enter a marathon. It is now very important that you follow the official route to the finish line or you will be disqualified. This would equate to operations such as milling the outside of a piece we are manufacturing, when the tool must follow an exact path. This process is called *continuous path control.* Obviously, it is more difficult to control an exact path. All CNC machines today use the continuous path method.

Machine Tool Axes

CNC machines can also be classified on the number of axes, or directions of motion, that they are capable of. Machining centers usually have two-and-a-half, three, four, or five axes with three and four axes being the most common. The four and five axes machines incorporate a rotary table of some sort to be capable of continuous motion in all axes simultaneously.

Lathes and turning centers generally have between two and four axes. The standard configuration consists of a two-axis lathe with one axis parallel to the spindle and one axis perpendicular to the spindle. A four-axis turning center has a rear or second turret that can be programmed independently of the master turret.

Think about a vertical milling machine. The X axis is the table movement right and left as you face the machine. The Y axis is the table movement toward and away from you. The Z axis is the spindle movement up and down. A move toward the work is a negative Z (−Z) move. A move up in this axis would be a positive Z (+Z) move.

Milling machines use all three axes, as seen in Figure 1–6. The X axis usually has the longest travel. On a common vertical milling machine, the X axis moves to the operator's left and right. The Y axis moves toward and away from the operator. The Y axis usually has the shortest travel. The Z axis always denotes movement parallel to the spindle axis, the up and down movement. Toward the work is a negative Z move. It may be helpful to think of the motion in terms of tool position. If the table is moved to the left, the tool is positioned more in the + X direction. If the table is moved toward the front, the tool is positioned more in the −Y direction.

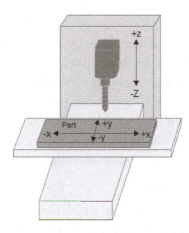

FIGURE 1–6. Typical milling machine configuration illustrating the X, Y, and Z

Components of CNC Machines

CNC Controls

Machine controls have changed greatly with the development of the computer. Many of the machine's operating characteristics can be changed by the operator so that the machine operates the way they want it to. Parameter tables in CNC machines today allow each machine to be personalized to the needs of the job to be run.

Computers allow CNC machines to store incredible amounts of data and programs. They enable the use of inexpensive computer hardware such as thumb drives and networking.

Figure 1–7 shows PC-based mill and turning center controller graphics.

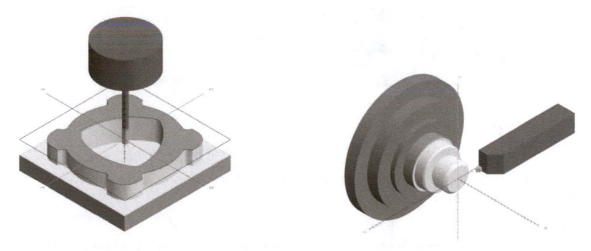

FIGURE 1–7. PC-based mill graphics and PC-based turning center graphics.

Many CNC controllers also allow the user to add programming software directly to the machine. For example, one could install Gibbs programming software on a PC-based controller and use it to develop programs for the machine.

Displays

Displays allow the operator easy access to information. On the screen of the machine tool, the operator can see the program, tool and cutter offsets, machine positions, variables, alarms, error messages, spindle RPM, and horsepower usage. Most also have tool-path simulation. The control reads the program and simulates the tool path on the screen. The tool-path simulation can be used to prove the program before the program is run to eliminate programming errors that could damage tools, parts, machine, or operator. The simulation can also be used to approximate the machining time. Many machines also have maintenance and troubleshooting information available on the display.

Servo Motors and Closed-Loop Systems

AC or DC servo motors are used to turn ball screws, which in turn drive the different axes of the machine tool. Servo motors permit untended operation, and closed-loop control makes sure that the machine actually does what the program told it to do (see Figure 1–8). Computers make it possible to continuously monitor a machine's position and velocity while it is operating. For example, if we tell the machine to move 10 inches at a feed rate of 5 inches per minute, the computer will constantly monitor the axis to be sure it is properly executing the move. If the table were to run into an obstruction, the computer would know instantly that it should be moving but it is not. The computer would then stop the motion and signal an error condition. This usually occurs before serious damage is done to the machine.

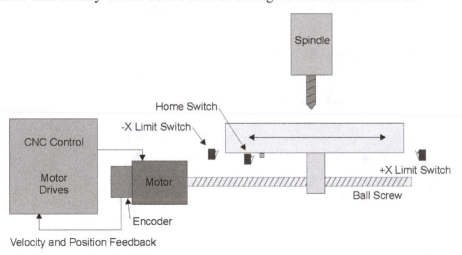

FIGURE 1–8. Closed-loop system configuration.

The advantages of the servo motor include motor and feedback mechanism, increased travel and spindle speeds, and increased accuracy and repeatability.

Homing

Some CNC machines need to be homed. Homing moves each axis to a known position where each axis is initialized. It is crucial that a machine home at the exact same position each time so that programs and fixture locations can be accurate and consistent.

The procedure is very simple. The operator moves each axis to a safe location and then hits a home key for each axis. Safe location means that there are no fixtures, parts, or other tooling that the machine will run into as it homes.

Study Figure 1–8 to see what happens when one axis is homed. Note that there are three switches that the table can contact. There is a –X limit switch, a +X limit switch, and a home switch. The limit switches provide overtravel protection, which disables the motor drive if the table ever moves far enough to contact these switches. If we contact the –X limit switch, for example, the motor drive will be disabled for any further movement in the –X direction, and the control will go into an error condition. The operator will have to reset the error and move the axis in the +X direction until the table is off the –X limit switch. On many machines this will require the operator to hold an override key down while moving the table off the limit switch. The same is true if the +X switch is hit by the table. The operator would have to reset the error and move the table in the –X direction.

The home switch is used to roughly position the table during the homing routine. The operator moves the table to a safe position and hits the home key for the X axis (in this example). The table begins to move slowly toward the home switch. When the table contacts the switch, a signal is sent to the CNC control, which senses that the switch has closed.

The CNC control then reverses the motor and very slowly turns it. The CNC control monitors the encoder until it sees the home pulse. When the CNC control sees the home pulse from the encoder, it initializes its position. It now knows exactly where this axis is. The home switch assures that the table is close to position and that we are in the right revolution of the encoder. The encoder home pulse establishes the home position exactly. Each axis homes in the same manner.

Note that most machines also have software axis limits stored in computer memory. For example, there may be a limit of –12 inches for the X axis. If the X axis ever gets to a –12 inch position, the CNC control stops the –X axis travel and turns on an error code. The software limits are set up to be active before the actual limit switches are hit, providing extra protection. Many newer CNC machines do not need to be homed. They retain their position information through absolute encoders.

Ball Screws

The rotary motion generated by the drive motors is converted to linear motion by recirculating ball screws. The ball lead screw uses rolling motion rather than the sliding motion of a normal lead screw. Sliding motion is used on conventional Acme lead screws. Acme lead screws work on a friction and backlash principle (see Figure 1-9); ball screws do not (see Figure 1–10). Acme lead screws were used on conventional machine tools. Note in Figure 1-10 that every time the travel reverses backlash occurs.

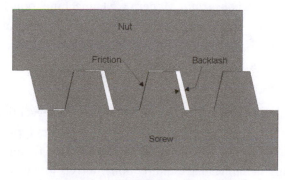

FIGURE 1–9. Friction and backlash are two disadvantages of the conventional Acme lead screw.

The balls, located inside of the ball screw nut, contact the hardened and ground lead screw and recirculate in and out of the thread (see Figure 1–10). The contact points of the ball and screw directly oppose one another and virtually eliminate backlash. The contact points are also very small, so very little friction is generated between them.

Other advantages of the ball lead screw over the Acme lead screw are:
1. less wear
2. high speed capability
3. precise position and repeatability
4. longer life

FIGURE 1–10. Recirculating ball screws have almost no backlash.

Backlash Compensation

Note that although ball screws greatly reduce backlash, there is no such thing as zero backlash. Machine tool manufacturers have incorporated an electronic pitch error or backlash compensator into most CNC machines. The backlash compensator corrects errors detected by lasers at the time the machine is assembled. The amount of compensation is loaded into a storage area within the control by the machine-tool representative. As the machine moves, the control adjusts the position of the machine according to the stored data. This number can be changed in memory as the machine wears.

The CNC Machine

The machine tool itself has evolved along with the computer control and drive systems. In the beginning, the majority of numerical control machines were just conventional machine tools with a control added on to the manual machine. This is called a *retrofit*. Modern CNC machine tools have been completely redesigned for CNC machining and bear little resemblance to their conventional counterparts. Requirements of new machine tools include rigidity, rapid mechanical response, low inertia of moving parts, high accuracy, and low friction along the ways.

Tool Changing

Tool changers automatically change cutting tools without operator intervention. Machining centers can have anywhere from 16 to 100 tooling stations. Lathe or turning centers typically have 8–12 tools, which are indexed automatically. Quickness of tool changing is an important factor on machines used in production environments. Many machines have bidirectional turrets that can select the quickest route to the desired tool.

Programming and Input Media

Programming may be done on the CNC control or on a remote computer. Programming on the machine is usually done with conversational programming. Conversation programming enables the programmer to develop the program by describing the geometry of the part. In many CNC systems, the conversational input is changed to G- and M-code programs). This is why programmers and operators need a strong background in G-code language. The software on a remote computer is often called computer-aided manufacturing (CAM) software. CAM software allows the programmer to describe the part geometry. The computer then post-processes the CAM part file into a CNC control program for a particular machine. Then the program is loaded into the CNC machine with a USB drive or over the network.

Why CNC?

Consider a typical job (see Figure 1–11). An operator would study the blueprint and plan the correct machining methods and sequence. Next, the operator would gather the tools necessary to do the job. In this example, there is a fixture for milling and a fixture for drilling.

The operator would gather the tools to complete the job. The operator then finds an available milling machine and properly locates and fastens the milling fixture onto the table.

After loading the end mill into the spindle and adjusting the speeds and feeds of the machine, the operator turns on the coolant and spindle. The operator then very carefully moves the table handles to move the table and spindle to the proper locations to mill the part.

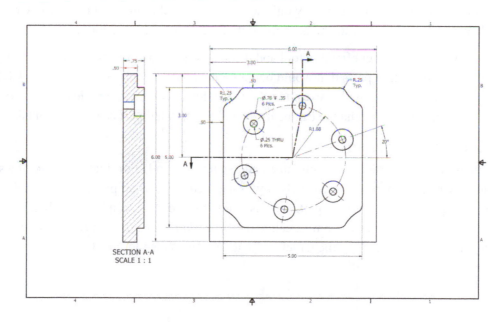

FIGURE 1–11. This typical job shop part could be done in one setup using CNC machine tools.

When the milling is complete, the operator chooses a drill press. The operator carefully mounts the drilling fixture on the table and loads the drilling tools. Proper speeds and feeds are set, and the coolant and spindle are turned on. The operator then guides the drill into the fixture to correctly locate the holes to be machined.

Contrast how this job is done on the CNC machine. The operator is given the job. The operator downloads the part program into the CNC machining center and locates the first piece in the vise. After using an edge finder to locate the piece, the operator runs the program. The machine turns on the coolant, loads the proper tools, sets the proper speeds and feeds, and machines the part while the operator watches.

We do not need drill fixtures to guide the drills to assure accurate location; the machine drills the holes precisely on location. In many cases, the operator can just clamp the part in the vise on the machine. This helps reduce the cost of producing a part because we do not have to design and build a fixture.

It helps reduce lead time because we can make the part right away; we do not have to wait until a fixture is built. Lead time is the time between when we receive an order and when we can make and ship it.

If our customer wants a change in a dimension of the part or a hole moved, we can do it immediately with a CNC machine. We would not have to rework the fixture. Set-up time is dramatically reduced with a CNC machine. Remember that a machine is only making money while it is actually machining. We are not making money while the operator loads tools, lines up and clamps fixtures on the table, and so on.

CNC machines dramatically reduce the non-machining time. The CNC can make very rapid positioning moves between machining operations. Many CNC machines have tool changers and tool carousels that can hold many tools. Most of the common tools will already be available in the CNC so the operator doesn't have to load tools. The CNC can change a tool in seconds, while an operator might take minutes. Remember that the machine is making money only while actual machining is taking place. CNC helps drastically reduce non-machining time and is appropriate for small- and large-lot production.

CNC machines have been very beneficial in tool and die shops. Skilled craftsmen can use the machines to produce complex, one-of-a-kind parts that would be very time consuming if done manually. CNC machines are very effective for large lot sizes, too. CNC machines also enable rapid changes to parts. If a change is needed a quick program change can be made and the updated part can be machined.

Axes and Coordinate Systems

To fully understand CNC programming, you must understand axes and coordinates. Think of a simple part. You could describe the part to someone else by its geometry. For example, the part is a 4-inch by 6-inch rectangle. Any point on a machined part, such as a hole to be drilled, can be described in terms of its position. The system that allows us to do this, called the *Cartesian coordinate* or *rectangular coordinate system,* was developed by a French mathematician, René Descartes (see Figure 1–12).

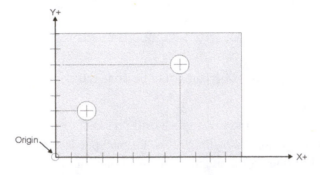

FIGURE 1–12. An example of coordinate positioning.

The Cartesian Coordinate System

Consider just one axis first. Imagine a line with zero marked in the center (see Figure 1–12). Now imagine that to the right of zero we have marked every inch with a positive number representing how far the mark is to the right of zero. Mark the inches to the left of zero beginning with −1, −2, −3 and so on. This line is called the X axis. We could describe a particular position on our line by giving its position in inches from zero. This would be called a position's *coordinate*.

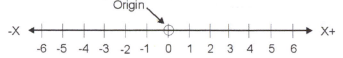

FIGURE 1–13. Single-axis coordinate line.

Let's add a perpendicular line (axis) that crosses our first line (X axis) at zero. The horizontal line is the X axis and the vertical line is the Y axis. The point where the lines cross is the zero point, usually called the *origin*. Points are described by their distance along the axis and by their direction from the origin by a plus (+) or minus (−) sign (see Figure 1–14).

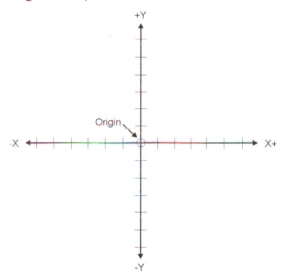

FIGURE 1–14. Dual-axis coordinate grid.

Quadrants

The axes divide the work envelope into four sections called *quadrants* (see Figure 1–15). The quadrants are numbered in a counterclockwise direction, starting from the upper right.

Points in the upper right, quadrant 1, have positive X (+X) and positive Y (+Y) values.

Points in quadrant 2 have negative X (−X) and positive Y (+Y) values.

Points in quadrant 3 have negative X (−X) and negative Y (−Y) values.

Points in quadrant 4 have positive X (+X) and negative Y (−Y) values.

13

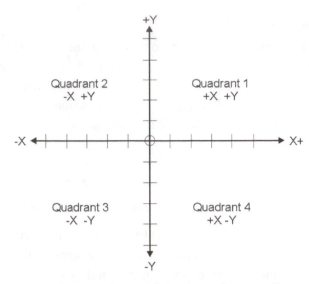

FIGURE 1–15. The four quadrants of the Cartesian coordinate system.

To locate a point such as (X3.0, Y3.0) in the two-axis system, start at the zero point and count to the right (+ move) three units on the X axis and up (+ move) three units on the Y axis.

Figure 1–16 shows a point in the Cartesian coordinate system. The point's coordinates are identified as X3, Y3. Note that only one point would match these coordinates. Figure 1–17 shows another Cartesian coordinate system with eight points identified with their coordinates. Add one more axis (see Figure 1–18) to represent depth. If we are going to drill a hole, we would describe its location by its X and Y coordinates. We would use a Z value to represent the depth of the hole. The Z axis is added perpendicular to the X and Y axes.

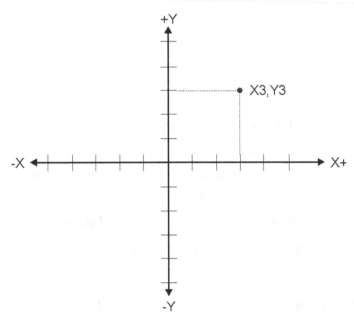

FIGURE 1–16. Locating a position using the Cartesian coordinate system.

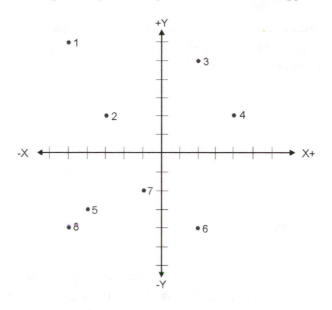

Location	X	Y
1	-5	6
2	-3	2
3	2	5
4	4	2
5	-4	-3
6	2	-4
7	-1	-2
8	-5	-4

FIGURE 1–17. Cartesian coordinate system and the XY coordinates for eight points.

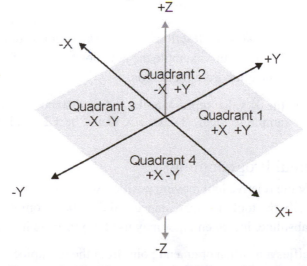

FIGURE 1–18. Three-axes coordinate grid.

Lathes or turning centers typically use only the X and Z axes. The Z denotes movement parallel to the spindle axis and controls the lengths of parts or shoulders. The X axis is perpendicular to the spindle and controls the diameters of the parts (see Figure 1–19).

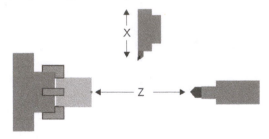

FIGURE 1–19. Typical lathe configuration illustrating the X and Z axes.

Polar Coordinates

It is also possible to describe the position of points by stating angles and distances along the angles. The direction of the angular line is viewed from the X axis. A positive angular dimension runs counterclockwise from your present position. A negative angular dimension would run clockwise from your present position. Study Figure 1-20. The desired position therefore is a point on the angular line, the desired radial distance from your present position (A+37, 1.250).

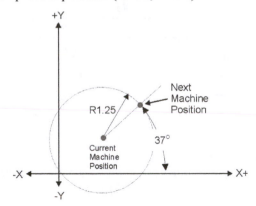

FIGURE 1–20. Polar coordinates are described from the current position, not from the absolute axes origin (X0,Y0). In this example the next position is 1.250" on a line that is +37⁰ through the current position from the X axis.

When using polar coordinates the X value represents the radius distance and the Y value represents the angle from 0 degrees. In the example in Figure 1-20 the X would be 1.25 and the Y would be 37 degrees in the polar coordinate mode.

Absolute and Incremental Programming

Absolute coordinates specify the relative tool moving position with respect to the program zero. Incremental coordinates specify the tool moving distance and direction from its current position. CNC programs can be written in absolute, incremental, or may use both formats in the same program.

Absolute programming specifies a position or an end point from the workpiece coordinate zero (datum). It is an absolute position (see Figure 1–21).

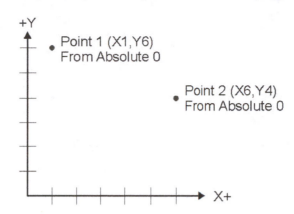

FIGURE 1–21. Absolute coordinate positions are always located from the program zero.

Incremental programming specifies the movement or distances from the point where you are currently located (see Figure 1–22). A move to the right or up from this position is always a positive move (+); a move to the left or down is always a negative move (–). With an incremental move, we are specifying how far and in what direction we want the machine to move.

Absolute positioning systems have a major advantage over incremental positioning. If the programmer makes a mistake when using absolute positioning, the mistake is isolated to the one location.

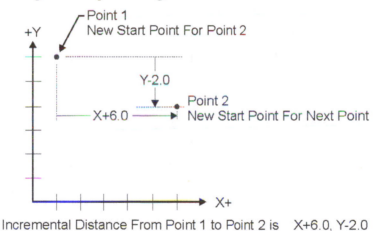

FIGURE 1–22. In incremental coordinate positioning, your present position becomes the program start position. From the start point, point 1 is one position to the right on the X axis (X+1) and six positions up on the Y axis (Y+6). When we move to point 2, point 1 becomes our start point. Point 2 is 5 positions to the right of our present position on the X axis (X+5) and 2 positions down on the Y axis (Y–2). Remember, in incremental programming, moves down or moves to the left are negative moves.

When the programmer makes a positioning error using incremental positioning, all future positions are affected. Most CNC machines allow the programmer to mix absolute and incremental programming. There are times when using both systems in one program will make the program easier to write.

CHAPTER QUESTIONS

1. Describe the 3 axes that a mill typically has.

2. What direction is negative Z in terms of the spindle?

3. Describe ball screws and explain why they are used on CNC machines.

4. Make a sketch of the four Cartesian quadrants and identify the signs of each quadrant.

5. Describe the incremental positioning mode.

6. Name one disadvantage of the incremental positioning mode.

7. Identify the positions marked on the Cartesian coordinate grid below. Use absolute positioning.

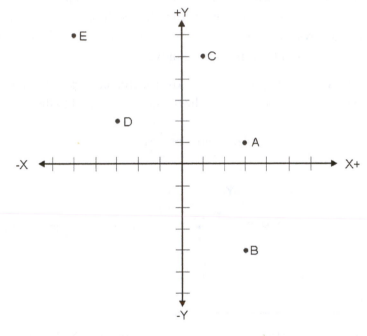

Point	X	Y
A		
B		
C		
D		
E		

Using the Cartesian coordinate system, write down the X and Y values for the incremental moves.

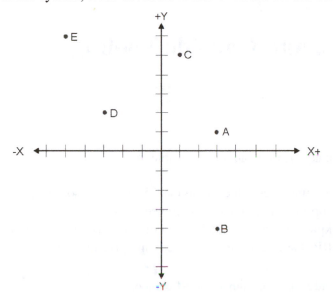

Point	X	Y
Origin to Point A		
Point A to Point B		
Point B to Point C		
Point C to Point D		
Point D to Point E		

Chapter 2

Speeds, Feeds, and Carbide Tooling

OBJECTIVES

Upon completion of this chapter, the reader will be able to:

- Define terms such as cutting speed, feed, chip load, SFPM, and so on.
- Locate cutting condition information using reference materials.
- Explain how the workpiece setup and machine rigidity affects cutting conditions.
- Calculate the proper RPM and feed rates for machining given the tool and material to be machined.
- State the two main characteristics of carbide.
- Describe the term "grade" as it applies to insert choice.
- Explain which factors to consider when selecting tool nose radius.
- Name three insert shapes in order of increasing strength.
- State two factors to consider when selecting insert shapes.
- State how insert size and depth of cut are interrelated.
- Describe the purpose and function of the different rake angles.
- Choose inserts and tool holders for various applications.
- Explain the purpose of qualified tools.
- Troubleshoot typical machining problems.

Overview

Speeds and feeds very important in machining. Improper speed and/or feed can cause excessive tool wear, heat, and/or tool breakage. The principles of cutting speeds and R.P.M. calculations apply to all machining processes. There are some special operations that require special speeds and feeds.

Cutting Speeds
Different materials need to be cut at different speeds. The term used to describe a material's ideal speed for cutting is cutting speed. Cutting speed is the speed at the outside edge of the cutter as it is rotating. This is also known as surface speed. Surface speed, surface footage, and surface area are all directly related.

Two wheels can illustrate an example of this (see Figure 2-1). Consider two wheels, one wheel is three feet in diameter and the other wheel is one foot in diameter. If each wheel is rolled one complete revolution the 3 foot wheel moves about 9 feet (3 * PI) and the 1 foot diameter wheel moves about 3 feet (1 * PI). The larger wheel traveled farther because it has a larger circumference.

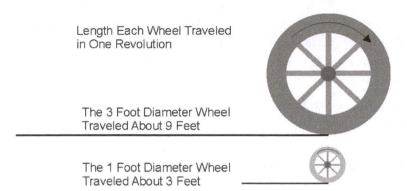

Length Each Wheel Traveled
in One Revolution

The 3 Foot Diameter Wheel
Traveled About 9 Feet

The 1 Foot Diameter Wheel
Traveled About 3 Feet

Figure 2-1. An illustration comparing the circumference of two different diameter wheels and cutting speed.

The same is true for cutting on a machine. Study Figure 2-2. This figure represents two different diameter parts being cut on a lathe. As the spindle of the lathe turns the part, the outside of the part (circumference) passes by the cutting tool. Loosely speaking, the cutter would make a 9-inch long chip on the 3 inch diameter part with each revolution of the spindle. In the bottom example the lathe would only make a 3-inch long chip in one revolution.

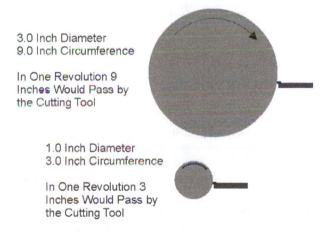

3.0 Inch Diameter
9.0 Inch Circumference

In One Revolution 9
Inches Would Pass by
the Cutting Tool

1.0 Inch Diameter
3.0 Inch Circumference

In One Revolution 3
Inches Would Pass by
the Cutting Tool

Figure 2-2. This figure shows the difference that diameter makes in cutting speed.

Next imagine different types of materials: plastic; aluminum; brass; mild steel; alloy steel; hardened steel, and so on. Each of these materials would have an ideal speed at which they should be machined. For example, if we are cutting alloy steel at too high a speed we will just burn up the tool. But that speed might have been fine for brass. So, every material has an ideal speed at which to be cut. These ideal speeds have been determined for all common materials. These ideal speeds are called cutting speeds. They can be found in cutting speed charts.

Cutting speeds are given in surface feet per minute (SFPM). Imagine that the parts in Figure 2-3 are made of mild steel. The cutting speed for mild steel with a HSS tool is 100 SFPM. This would mean the ideal cutting condition would be 100 feet of material passing the tool point every minute. Study Figure 2-3. The 3-inch diameter part has a 9 inch circumference. That means that 9 inches go by the tool with every revolution. 100 surface feet per minute would be 1200 inches per minute (100 * 12).

To get 1200 inches of material to pass the tool with a 9 inch circumference we would need to divide 1200 inches by 9 inches (1200" / 9"). This would give us 133 for an answer. The spindle would have to turn at 133 revolutions per minute (RPM) to equal 100 SFPM. This would be the ideal speed to cut mild steel with a 3 inch diameter.

How about the 1 inch diameter mild steel part? The ideal cutting speed is still 100 SFPM, but the circumference of the part is only 3 inches. So if we convert the 100 SFPM to inches we would have 1200 inches per minute. The part has a 3 inch circumference so 1200" / 3" = 400. The ideal spindle speed for this smaller diameter part would be 400 RPM (see Figure 2-3).

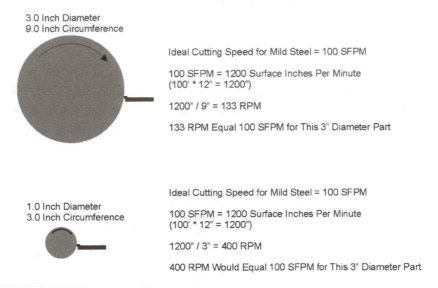

Figure 2-3. A comparison of machining a 3" part and a 1" part.

Note that in this example the smaller diameter required a higher RPM to achieve the ideal cutting speed for the material. The larger diameter part required a lower RPM to achieve the ideal cutting speed.

In fact the diameter of the larger part was 3 times as large as the 1 inch diameter part and the recommended RPM of the larger part was 1/3 the RPM of the smaller part.

How is the ideal cutting speed of a material determined? Cutting speeds are determined by the machinability rating. Machinability is the ability of a material to be machined. Machinability ratings determine recommended cutting speeds. Recommended cutting speeds are given in charts.

The machinability of the material determines the recommended cutting speed. The harder the material: the slower the cutting speed. The softer the work material the faster the recommended cutting speed (see Figure 2-4).

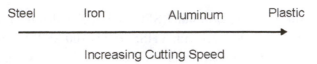

Figure 2-4. The softer the material: the higher the cutting speed.

Type of Cutting Tool and its Effect on Cutting Speed

The material the cutting tool is made of also determines the cutting speed. The two most common would be high-speed-steel (HSS) and carbide. Cutting speed tables will give a cutting speed for HSS tooling and also for carbide tooling. The cutting speed for carbide is usually about 4 times faster. If the cutting speed for a particular steel is 80 SFPM for a HSS end mill, the SFPM would be approximately 320 for the same material with carbide.

The harder the cutting tool material, the higher the cutting speed (see Figure 2-5). The softer the cutting tool material, the lower the recommended cutting speed.

Figure 2-5. The harder the material, the lower the cutting speed.

All cutting tools work on the surface footage principal. Cutting speeds depend primarily on the kind of material you are cutting and the kind of cutting tool you are using. Using the proper cutting speed for the material that is being machined will assure that cutting is optimal and is not too slow or fast for the tool.

Other Factors that Affect Cutting Speed

The depth of cut and the feed rate will also affect the cutting speed. These three factors: cutting speed; feed rate; and depth of cut are known as cutting conditions. Think of cutting conditions as being how much work is being done with the tool. A heavy roughing cut will create more heat and may require a lower cutting speed. While a light finish cut might allow a higher cutting speed.

Calculating RPM

The spindle RPM must be set so that the part or cutter will be operating at the correct cutting speed for the material being machined. In the previous examples we used a lathe part. The same considerations apply to a cutting tool for a mill or other machine. For the rest of the RPM calculation explanation we will talk about cutting tools, not a lathe part. But the same RPM calculations apply to both.

To determine the proper machining speed we need to calculate the revolutions per minute (RPM). The cutting speed or surface speed changes based on the size of the cutter. So to keep the surface speed correct for different sizes of cutters we must use a formula that includes the size of the cutter to achieve the proper surface footage.

The correct RPM will change with the size of the cutter. As the milling cutter gets smaller the RPM must increase to maintain the recommended cutting speed (SFPM). Remember the example of the wheel. Think of the cutter as a wheel and the cutting speed as a distance. A larger wheel (cutter) will need to turn fewer revolutions per minute to cover the same distance in the same amount of time than a smaller wheel (cutter). Therefore, to maintain the recommended cutting speed, larger diameter cutters must be run at slower speeds than smaller diameter cutters.

The machine must be set so that the cutter will be operating at the proper surface speed. Spindle speed settings on machines are in RPMs. To calculate the proper RPM for the tool, use the formula shown in Figure 2-6.

$$\frac{\text{Cutting Speed (CS)} * 4}{\text{Diameter of Cutter (D)}} = \text{RPM}$$

Figure 2-6. RPM formula.

A milling cut is to be taken with a 0.500 inch high speed steel (HSS) end mill on a piece of 1018 plain carbon steel. Calculate the RPM setting to perform this cut (see Figure 2-7). Use the recommended cutting speed charts in Figure 2-8.

Cutting Speed = 90 SFPM
Diamter of Cutter = 0.500

$$\frac{CS * 4}{D} = \frac{90 * 4}{0.500} = \frac{360}{0.500} = 720 \text{ RPM}$$

Figure 2-7. RPM calculation example.

Workpiece Material	Cutting Speed in SFPM	
	High-Speed Steel	Carbide
Plain Carbon Steel - 1018	90-120	270-450
6061-T6 Aluminum	70-100	225-375
4140	30-70	145-300
303	100-150	325-500
11L17 Leaded Free Machining	170	400
304/316 Stainless Steel	60	230
A2 Tool Steel	50	250

Figure 2-8. Recommended Cutting Speed for Milling in Surface Feet per Minute (SFPM)

If the machine does not have variable speeds, choose the speed which is nearest to the calculated RPM.

There are other considerations when choosing RPM. Are you roughing or making a finish cut? If you are roughing, go slower. If you are making a finish cut, go faster. What is your depth of cut? If it is a deep cut, go to the slower RPM setting. Go slower for setups that lack a great deal of rigidity. Are you using coolant? You may be able to go to the faster of the two settings if you are using coolant.

The greatest indicator of using the correct cutting speed is the color of the chip. When using a high-speed steel cutter the chips should never be turning brown or blue. Straw colored chips indicate that you are on the maximum edge of the cutting speed for your cutting conditions. When using carbide, chip colors can range from amber to blue, but should not be black. A dark purple color will indicate that you are on the maximum edge of your cutting conditions.

Let's try some examples.

A milling cut is to be taken with a 6.00 inch (HSS) face mill on a piece of 1045 steel (Cutting speed = 55 SFPM). Calculate the correct RPM for this cut (see Figure 2-9).

Cutting Speed = 55 SFPM
Diameter of Cutter = 6.00

$$\frac{CS * 4}{D} = \frac{55 * 4}{6.00} = \frac{220}{6.00} = 36 \text{ RPM}$$

Figure 2-9. RPM Calculation example.

A 1-inch (HSS) drill is used on a piece of 1018 steel with a cutting speed of 100 SFPM. Calculate the RPM for this drilling operation (see Figure 2-10).

Cutting Speed = 100 SFPM
Diameter of Cutter = 1.00

$$\frac{CS * 4}{D} = \frac{100 * 4}{1.00} = \frac{400}{1.00} = 400 \text{ RPM}$$

Figure 2-10. RPM Calculation example.

A cut is to be taken with a 3.00 inch carbide face mill on a piece of 4140 alloy steel with a cutting speed of 300 SFPM. Calculate the RPM for this cut (see Figure 2-11).

Cutting Speed = 300 SFPM
Diameter of Cutter = 3.00

$$\frac{CS * 4}{D} = \frac{300 * 4}{3.00} = \frac{1200}{3.00} = 400 \text{ RPM}$$

Figure 2-11. RPM Calculation example.

Feed Rate Calculation

There are three factors that affect cutting conditions; cutting speed, depth of cut, and feed rate. The feed rate on milling machines is given in terms of inches per minute (IPM). IPM is the rate at which the tool will advance into the work.

The feed rate is determined by the RPM of the cutter, the number of cutting teeth on the cutter, and by the chip thickness that the cutter teeth can handle. The chip size is called the feed rate in inches per tooth or chip load.

Study Figure 2-12. This figure shows chip load for conventional milling and for climb milling. If we set a feed with a chip load of .005" it would mean that each tooth would take a .005" thick cut every revolution.

Take a close look at the climb milling example. Note the direction of table feed and the direction of tool rotation. In climb milling the cutter tends to "pull" the table in the same direction as the feed. Note also that in climb milling each tooth takes the full chip load as it encounters the material. Climb milling can produce better surface finishes if the machine is rigid and there is very little backlash in the table. If there is backlash or the machine is not rigid, chatter will result.

Next study the conventional milling example in the figure. The direction of the cutter rotation opposes the direction of the table feed. The chip load is very light as each tooth starts its cut. The thickness of the chip does not reach the full chip load until the end of the cut. Conventional milling will work better on machines that are not rigid or those with more backlash.

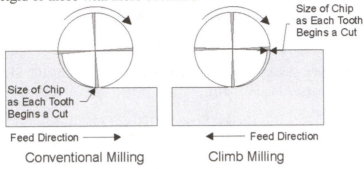

Figure 2-12. Conventional milling and climb milling.

The recommended values for chip load are based on the cutting tool material, the cutting tool size, and the hardness or machinability rating of the workpiece material. The recommended values for feed per tooth (chip load) can be found in charts, reference books, or on the internet. A typical feed in inches per tooth chart is shown in Figure 2-13.

While the recommended feed rates found in these charts represent good fundamental machining practice, they are only recommended values. Deviations from these recommended values may be necessary due to machining conditions. The feed rate used on small or thin work may need to be reduced. The work holding technique has a great deal to do with the feed rate. Setups, which lack rigidity, may require a slower feed rate. Feed rates on rigid milling machines can be much heavier than those feed rates used on lighter duty milling machines. When using large carbide face mills, the available horsepower and the rigidity of the spindle will always influence the feed rate.

Material	Chip Load for End Mills		Chip Load for HSS Drill
	HSS	Carbide	1/6 to ¾ Diameter
Mild Steel - 1018	.001 - .005	.015 - .006	.001 - .015
6061-T6 Aluminum	.002 - .006	.002 - .010	.001 - .016
4140 Steel	.001 - .004	.0015 - .006	.001 - .014
303 Stainless Steel	.001 - .005	.001 - .005	.001 - .014
11L17 Free Machining Steel	.001 - .005	.001 - .007	.001 - .018
304/316 Stainless Steel	.001 - .005	.0005 - .003	.001 - .010
A2 Tool Steel	.0005 - .003	.001 - .004	.001 - .007

Figure 2-13. Recommended feed in inches per tooth for high-speed steel milling cutters.

Feed Rate Calculation

The feed rate in inches per tooth (chip load) must be converted into feed rate in inches per minute (IPM) before you can set the feed rate on a machine. The formula for converting feed rate in inches per tooth into inches per minute is shown in Figure 2-14.

$$\text{Feed Rate (in./min.)} = \text{RPM} * \text{Chip Load} * \text{Number of Teeth}$$

Figure 2-14. Feed rate formula.

These feed rates are a starting point. You will typically be given a range of chip load factors to use. A good rule of thumb is to start out at the low range or average feed per tooth and increase the feed rate to the capacity of the machine tool, the setup, and the desired surface finish. It must also be mentioned that using a chip load that is too small will cause excessive tool wear.

Let's try some feed rate calculations. Follow along using the recommended feed rate charts in Figure 2-13. A four flute 0.5 inch high speed steel (HSS) end mill is to be used on a piece of 1018 mild steel with a cutting speed of 80. The RPM setting to perform this cut is 320 rpm. Look up the feed per tooth in the charts and calculate the feed rate in inches per minute (see Figure 2-15).

$$\frac{CS * 4}{D} = \frac{80 * 4}{0.50} = \frac{320}{0.50} = 640 \text{ RPM}$$

RPM = 640
Feed in Inches Per Tooth (Chip Load) = 0.005"
Number of Flutes = 2

Feed (Inches Per Minute) = RPM * Chip Load (CL) * Number of Teeth

Feed (Inches Per Minute) = 640 * 0.005 *2

Feed = 6.4 Inches Per Minute (IPM)

Figure 2-15. Feed rate calculation.

Some judgment must be used when selecting feed rates. The calculated feed rate is a good place to start. One must also consider the machining conditions. What are the surface finish requirements? A larger feed rate will leave a rougher finish. What is the depth of cut? If it is a deep cut, use a slower feed rate setting. Is the setup rigid? Go slower for setups that lack rigidity. Are you using coolant?

You may be able to have a higher feed if you are using coolant. So remember that the calculated feed rate is a good place to start. You may have to increase or decrease the feed rate to find the optimal feed rate.

Let's try another example. A two-flute .250 inch high speed steel (HSS) end mill is to be used on a piece of 8620 alloy steel with a cutting speed of 80 SFPM. Calculate the proper RPM. Also, calculate the feed rate in inches per minute using a chip load of .004 (study Figure 2-16).

$$\frac{CS * 4}{D} = \frac{80 * 4}{0.25} = \frac{320}{0.25} = 1280 \text{ RPM}$$

RPM = 1280
Feed in Inches Per Tooth (Chip Load) = 0.008"
Number of Flutes = 2

Feed (Inches Per Minute) = RPM * Chip Load (CL) * Number of Teeth

Feed (Inches Per Minute) = 1280 * 0.004 *2

Feed = 10.24 Inches Per Minute (IPM)

Figure 2-16. Feed rate calculation.

Depth of Cut

The depth of cut also affects speeds and feeds. The deeper the cut, the more work is being done. More work creates more heat in the tool and workpiece. Excessive heat can destroy tools. Therefore, to determine the depth of cut we must first select the proper cutting tool, the proper machine, and a suitable setup. There is a term called the metal removal rate. Metal removal rate is the number of cubic-inches-per minute of material that can be removed from the part.

Tool Type

Roughing tools should be used when large amounts of material need to be removed. The serrated flute design enables a roughing end mill (see Figure 2-17) to remove three times as much material as a plain end mill.

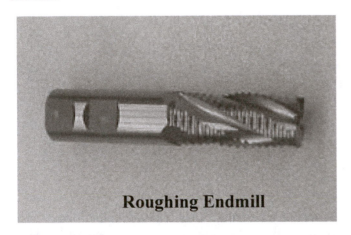

Roughing Endmill **Plain Endmill**

Figure 2-17. Roughing end mills can dramatically increase metal removal rates.

You should always use the largest tool that will perform the job. Larger tools are capable of larger metal removal rates. In some cases more than one tool will need to be used. It may mean changing tools, one tool for roughing cuts and one tool for finish cuts. The profile of the part and efficiency may require that you need to rough with a larger cutting tool and finish with a smaller tool.

Selecting the Proper Machine

The machine type and size will influence the depth of cut. How rigid is the machine? The easiest way to determine the rigidity of the machine is to look at the size of the spindle taper. The larger the spindle taper, the greater the rigidity of the machine. Make sure your metal removal rate decisions account for the style and size of the machine.

Available Horsepower

Does the machine have enough horsepower for the desired metal removal rate with the cutting tool and conditions? Roughing cuts with carbide on tough materials may require high horsepower.

The horsepower that is available at the spindle is not the same as the horsepower of the main motor. The available horsepower at the spindle may be as low as 50 to 80 percent of the main motor horsepower. For maximum metal removal rates you may need to calculate horsepower requirements. There are charts that show the required horsepower for various metal removal rates.

Rigidity of the Setup

How the part is clamped to the table and how far the tool is extended from the spindle are the major considerations in the setup. If the tool is extended and the workpiece can only be clamped in a limited number of places, the depth of cut and the feed rate may have to be reduced.

The more rigid the setup is, the higher the metal removal rate can be. The shorter the tool, the higher the metal removal rate can be. The rigidity of the setup should always be maximized for productivity and safety.

Depth of Cut

When using a plain high-speed-steel end mill the rule of thumb is the depth of cut should not exceed 1/2 the diameter of the cutter. An operator should maximize the feed rate and the depth of cut while keeping the RPM in the calculated range. When maximizing the depth of cut, use a slower feed rate with an acceptable chip load factor.

If the feed rate needs to be less than the recommended chip load, decrease the depth of cut instead. A feed rate that is too low will prematurely dull the end mill. Milling cutters with fewer teeth will allow a greater depth of cut. Fewer teeth make each tooth larger and stronger. Fewer teeth also allow more chip clearance. Use cutting fluids whenever possible. Cutting fluids dissipate heat. When a smooth, accurate finish is needed, take a roughing cut first to remove the majority of the material and then a finish cut.

Fundamentals of Carbide Tooling

Tooling is crucial to the success of a machine shop. A great machine with poor tooling will perform very poorly. Because good tooling techniques are vital to productivity, the rest of this chapter will concentrate on the fundamentals of carbide cutting tools.

Cemented Carbide

Cemented carbide, or tungsten carbide, is a form of powdered metallurgy. Fine powders consisting of tungsten carbide and other hard metals bonded with cobalt are pressed into required shapes and then sintered.

Sintering is the heating of the carbide materials to approximately 2,500 degrees Fahrenheit. At this temperature, the cobalt melts and flows around the carbide materials. Cobalt acts as the binder that holds the carbide particles together. After the carbide insert cools, the insert is almost as hard as a diamond.

The hardness and physical properties of cemented carbides enable them to operate at high cutting speeds and feeds. The great hardness of carbide is also its Achilles' heel. Extremely hard materials are also very brittle, and this can cause problems under certain machining conditions. Through the use of different mixtures of materials, carbide manufacturers have come up with different grades of carbide materials. Selecting the proper grade for the machining application is important for economy and productivity.

Selection of Carbide Tool Grade

Carbide tools come in a variety of grades. The grade is based on the carbide's wear resistance and toughness. As an insert becomes harder or more wear resistant, it becomes brittle (less tough).

If you use a very hard, wear-resistant insert on a material that has an uneven or interrupted surface (interrupted cut), the insert will most likely break. The ANSI/ISO standards organizations have devised systems of grading carbide based on the carbide insert's application and physical makeup (see Figure 2-18).

Classification systems differ and can be quite confusing. There are over 5000 different grades of carbide under 1500 trade names. Carbide manufacturers help clarify matters with cross-reference charts in their catalogues (see Figure 2–19).

ISO Grade	Material to be Machined
P01	Steel, Steel Castings
P10	Steel, Steel Castings
P20	Steel, Steel Castings, Ductile Cast Iron with Long Chips
P30	Steel, Steel Castings, Ductile Cast Iron with Long Chips
P40	Steel, Steel Castings with Sand Inclusions and Cavities
P50	Steel, Steel Castings of Medium or Low Tensile Strength, with Sand
M10	Steel, Steel Castings, Manganese Steel, Gray Cast Iron, Alloy Cast Iron
M20	Steel, Steel Castings, Austenitic Steel or Manganese Steel, Gray Cast Iron
M30	Steel, Steel Castings, Austenitic Steel or Manganese Steel, Gray Cast
M40	Mild, Free Cutting Steel, Low-Tensile Steel, Non-Ferrous Metals and Light
K01	Very Hard Cast Iron, Chilled Castings over 85 Shore Hardness, High
K10	Gray Cast Iron Over 220 Brinell Hardness, Malleable Cast Iron with Short
K20	Gray Cast Iron over 220 Brinell Hardness, Non-Ferrous Metals, Copper,
K30	Low-Tensile Gray Cast Iron, Low-Tensile Steel, Compressed Wood
K40	Soft Wood or Hard Wood, Non-Ferrous Metals

Figure 2-18. ISO Grade Designations

Coatings for Carbide Inserts

Carbide is a very hard, durable cutting tool, but it still wears. The wear resistance of cemented carbide can be greatly increased by using coated carbide inserts.

Wear-resistant coatings can be applied to the carbide substrate (base material) through the use of plasma coating or vapor deposition. The coating is very thin but very hard.

The most common types of coatings include titanium carbide (TiC), titanium nitride (TiN), and aluminum oxide (AlO). Aluminum oxide is a very wear-resistant coating used in high-speed finish cuts and light roughing cuts on most steels and all cast irons. Titanium nitride coatings are very hard and have the strength characteristics to perform well under heavy rough-cutting conditions. All three coatings will perform well on most steels, as well as on cast iron.

Diamond-Coated Inserts

Coated cutting tools have been around for years. Titanium and boron nitride materials have driven the coated cutting tool industry. One of the newer materials is the polycrystalline diamond, or PCD. PCD tools are becoming widely accepted as tooling solutions for difficult-to-machine materials. PCD material has the hardness of a diamond and the friction coefficient of Teflon. This combination has resulted in a remarkable increase in tool life.

ISO/ANSI Grade	Valenite	Iscar	Sandvik	Mitsubishi	Kennametal	Walter	Seco	Toshiba
P50-P40 C5	VP5535	IC635	GC4235		KC8050 KC9040 KC9240 KC9140			
P40-P30 C5-C6	VP5535 VPUP30	IC9025 IC3028 IC635	GC4235	UE6035	KC9040 KC8050 TN7035 KC5025	WPP30 WAP30	TP400 TP3000	T9035 T9025 TD930
P30-P20 C6-C7	VP5525	IC9025 IC3028 IC9015 IC50M	GC4225	UE6020 UE6035 UP20M VP15TF	KC9125 TN7025 KC8050 KC5025	WPP20 WAP30 WAP20	TP2500 TP2000 CP500	T9025 TD930 T7020 T725X
P20-P10 C5	VP5515	IC8048 IC570 IC9015 IC907	GC4215 GC1525	UE6010 UE6110 AP20N UP35N	KC9110 KC9010 TN7010 KC5010	WPP10 WAP20 WAP10	TP2000 TP1000 TP200	T9015 AT530 T7010 T715X
P10-P01 C8	VP1510 VP1505 VPUP10	IC9015 IC428 IC520N IC8048	GC4005 CT5015	UE6005 UE6010 AP25N	KC9110 KC9010 TN7005 KC9315	WPP01 WAP10 WAP01 WPP05	TP1000 TP100	T9005 TD905 AT520
M40-M30	VP5535 VP9625 VPUS10	IC9025 IC3028 IC635	GC2035	US735	KC9240 KC9245 CL4 KC8050 TN7035	WSM30 WAM30	TP400 CP500	T6030 T725X J740
M30-M20	VP8525 VP9625 VPUS10	IC3028 IC9025 IC08	GC2025	US7020 UP20M VP15F	KC9225 KC8050 KC5020 TN7025 TN8025	WAM20	TP3000 TP2500 CP500 CP200	T6030 T6020 AH120
M20-M10	VP8515 VP9610 VPUS10	IC907 IC570 IC507 IC520	GC2015 GC1025	US7020 UP20M AP25N	KC5010 KT315 KC5510	WAM10 WXN10	TP100 TP200 CP200	T6020 J530

FIGURE 2–19. Typical grade system cross reference chart gives grade selection choices for different tool manufacturers.

Insert Classification

Insert Shape

Indexable inserts come in many shapes. Inserts are clamped in tool holders and provide cutting tools with multiple cutting edges. After the cutting edges are worn to a point where they can no longer be used, they are discarded or saved and recycled. To correctly select the insert for turning, the machinist must be able to answer a few specific questions about the job.

- What geometric features are required on the workpiece?
- What are the characteristics of the material that will be cut?
- How much material is being removed?
- How rigid is the work holding setup?

31

Tool selection and identification of carbide tools is done using a standard set of identifying letters and numbers known as ANSI standard tool nomenclature. An example of a typical insert style, shape, and size is a CNMG432. The first identifying feature found on all carbide inserts is shape. Insert shape is dictated by the part shape required. Position 1 on the chart identifies the shape of the insert (see Figure 2-20). The C in this example would mean that we have an 80 degree diamond shaped insert.

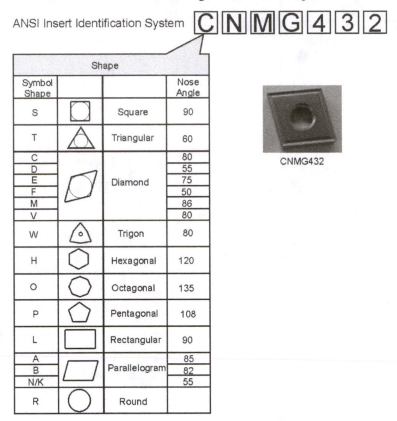

ANSI Insert Identification System C N M G 4 3 2

Shape			
Symbol Shape			Nose Angle
S	Square	Square	90
T	Triangular	Triangular	60
C	Diamond	Diamond	80
D			55
E			75
F			50
M			86
V			80
W	Trigon	Trigon	80
H	Hexagonal	Hexagonal	120
O	Octagonal	Octagonal	135
P	Pentagonal	Pentagonal	108
L	Rectangular	Rectangular	90
A	Parallelogram	Parallelogram	85
B			82
N/K			55
R	Round	Round	

CNMG432

Figure 2-20. Insert Shape.

Figure 2–21 shows the characteristics of different-shaped inserts in order of their strength. Round inserts have the greatest strength, as well as the greatest number of cutting edges, but the round configuration limits the operations that can be performed.

Square or 90-degree inserts have less strength and fewer cutting edges than round inserts but are a little more versatile.

Triangular inserts (see Figure 2–21) are more versatile than square inserts, but as the included angle is reduced from 90 degrees to 60 degrees it becomes weaker and more likely to break under heavy machining conditions.

Diamond-shaped inserts are probably the most commonly used shape. Diamond-shaped inserts range from a 35 degree to an 80 degree included angle. Diamond-shaped inserts are much more versatile than square and round inserts. It is good machining practice to select the largest included angle insert that will cut the shape of the part because the insert will be stronger.

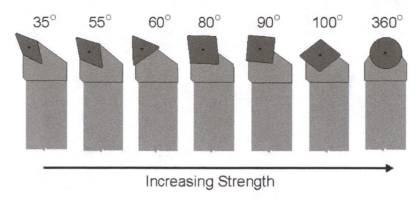

Increasing Strength

FIGURE 2–21. The shape of the insert will have a great effect on the strength of the tool. Select the largest included angle that will cut the part.

Lead Angle

Lead angle, or side-cutting edge angle, is the angle at which the cutting tool enters the work (see Figure 2–22). The lead angle can be positive, neutral, or negative. The tool holder always dictates the amount of lead angle a tool will have.

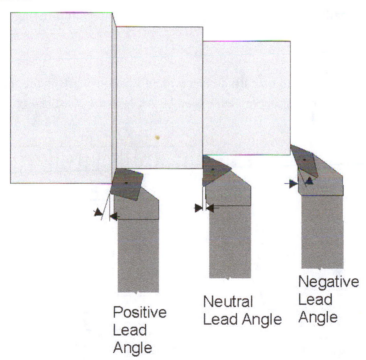

FIGURE 2–22. Lead or side-cutting edge angle is determined by the tool holder type. The lead angle can be positive, neutral, or negative.

Tool holders should be selected to provide the greatest amount of lead angle that the job will permit. There are two advantages to using a large lead angle. First, when the tool initially enters the work, it is at the middle of the insert where it is strongest, instead of at the tool tip, which is the weakest point of the tool. Second, the cutting forces are spread over a wider area, reducing the chip thickness (see Figure 2–23).

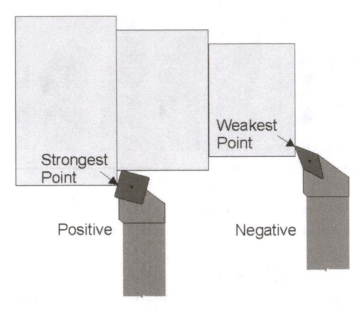

FIGURE 2–23. The effect of the lead angle on the strength of the insert. Increasing the lead angle will greatly reduce tool breakage when roughing or cutting interrupted surfaces.

Clearance Relief Angle

The second identifying feature found on carbide insert identification charts is the relief angle (see Figure 2-24). For our CNMG432 insert the second letter is an N. An N in the chart in Figure 2-24 would be a 0 degree relief angle.

Figure 2-24. Relief Angle

The side relief angle, also known as the side rake angle, is formed by the top face of the cutting tool and side cutting edge (see Figure 2-25). The angle is measured in the amount of relief under the cutting edge.

Top View

Side View

Side Relief Angle

Figure 2-25. Relief angle

Neutral rake inserts have an included angle of 90 degrees between the top rake and the end clearance angle (see Figure 2-26).

Top view

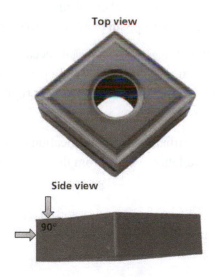

Side view

90°

Figure 2-26. Neutral Rake Insert.

This creates a 0 relief condition. 0 relief inserts have a letter designation of N or neutral (see Figure 2-24). Relief under the cutting edge is essential. Negative rake holders must be used when using neutral rake inserts. The combination of the relief angle on the insert and the rake angle created by the holder is known as the effective rake angle. There are three principal rake angles: neutral, positive, and negative (see Figure 2-27).

Neutral Rake   Positive Rake   Negative Rake

FIGURE 2–27. Side view of back rake angles.

When selecting the proper rake, it is essential to look at the machining conditions. Negative rake holders are a good, economical choice because they hold neutral rake inserts. Neutral rake inserts have twice as many cutting edges as positive rake inserts because the insert can be turned over and used.

Another advantage is that negative rake tool holders provide more support for the cutting edges of the insert. Under normal operating conditions, negative rake inserts are also a little stronger because of the compressive strength of carbide. Negative rake holders should be used when the tool and the work are held very rigidly and when high machining speeds and feeds can be maintained. More horsepower is required to cut with negative rake tool holders, which is why there is an increasing trend toward the use of positive rake cutting.

Positive rake cutting is more of a shearing effect than the pushing effect generated by negative rake. Positive rake holders generate less cutting force and have less of a tendency to chatter. Horsepower requirements are greatly reduced with positive cutting tools.

The only drawback to positive rake cutting tools is their inability to stand up to harder materials. Recent advances in carbide technology have produced tougher substrate materials that provide greater edge strength. Some carbide companies recommend positive rake holders whenever possible.

Positive rake should be used when machining softer materials because the chips are able to flow away from the cutting edge freely and the cutting action is more of a peeling effect. Positive rake cutting can be very successful on long slender parts or other operations that lack rigidity.

Insert Size Tolerance
The third identifying feature found on carbide insert identification charts is the insert size tolerance (see Figure 2-28). For our example the third letter is a M. From the chart you will see that there are three tolerances specified for an M.

This letter designation states how much size variation is allowed from one insert to the next. The tolerance that is described by this letter designation includes the nose radius, the insert thickness, and the inscribed circle.

Hole and Chip Breaker Configuration
The fourth identifying feature on the ANSI insert identification chart designates the shape of the chip breaker that is molded into the insert and whether the insert has a hole (see Figure 2-29). Our insert example is a G.

If the insert has a hole, the insert holder uses a lock pin to locate the insert. If there is no letter designation in the fourth field of the chart, then the insert doesn't have a hole. Inserts without holes are held only by a clamp on the holder.

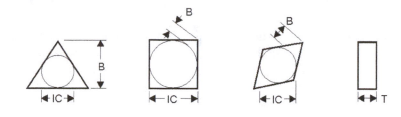

Tolerance Class	Tolerance on "B"		Tolerance on "IC"		Tolerance on "T"	
	INCH	MM	INCH	MM	INCH	MM
A	±.0002	± .005	±.001	±.025	±.001	±.025
C	±.0005	± .013	±.001	±.025	±.001	±.025
E	± .001	± .025	±.001	±.025	±.001	±.025
F	±.0002	±.005	±.0005	±.025	±.001	±.025
G	±.001	±.025	±.001	±.13	±.005	±.13
H	±.0005	.013	±.0005	±.025	±.001	±.025
J	± .002	±.005	±.002-.005	±.025	±.001	±.025
K	±.0005	±.013	±.002-.005	±.025	±.001	±.025
L	± .001	± .025	±.002-.005	±.025	±.001	±.025
M	±.002-.005	±.05-.13	±.002-.005	±.13	±.005	±.025
U	± .005-012	±.06-.25	±.005-.010	±.13	±.005	±.13
	Tolerance					

CNMG432

ANSI Insert Identification System   C N M G 4 3 2

FIGURE 2–28. Insert Size Tolerance.

ANSI Insert Identification System   C N M G 4 3 2

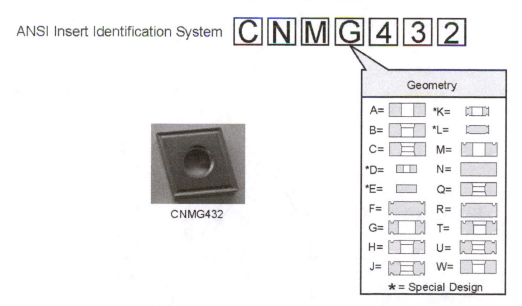

CNMG432

Figure 2-29. Hole and chip breaker configuration.

Chip Breaker

To select the proper chip breaking geometry, the depth of cut and the feed rates must be taken into consideration. If a roughing insert chip breaker is selected, the proper feed rate and depth of cut must be used. Using too light a feed rate or too small a depth of cut will result in long stringy chip. An insert chip breaker that is designed for finish cuts will fail if it is used with a large depth of cut or heavy feed. Chip breaker configuration is covered later in this chapter.

Insert Size

The fifth identifying feature on the ANSI insert identification chart designates the size of the insert (see Figure 2-30). Our example has a 4 in the 5th position. This would mean that the inscribed circle would be .5" (4*1/8).

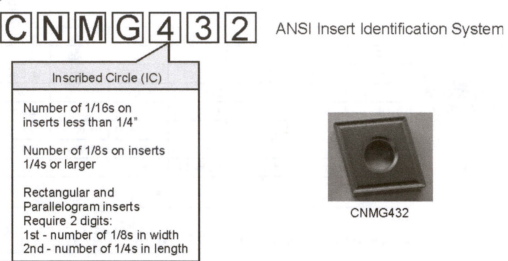

Figure 2-30. Insert Size.

The size of the insert is based on the inscribed circle (IC, which is the largest circle that will fit inside the insert), the insert thickness, and the tool-nose radius (see Figure 2–31).

The depth of cut possible with an insert depends greatly on the insert size. The depth of cut should always be as great as the conditions will allow. A good rule of thumb is to select an insert with an inscribed circle at least twice that of the depth of cut.

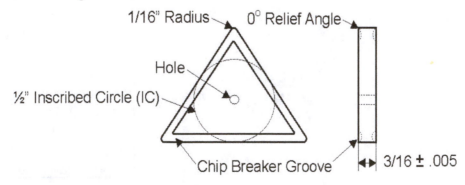

FIGURE 2–31. Diagram of a typical triangular insert.

Insert Thickness

The number in the sixth position of the ANSI Insert identification system chart represents the number of sixteenths of thickness. For our example we have a three in the sixth position. That would mean that the insert is 3/16 of an inch thick (see Figure 2-32). As the thickness of the insert increases, so does its strength. Remember, larger inserts can take deeper cuts. Insert thickness and inscribed circle size increase and decrease proportionally.

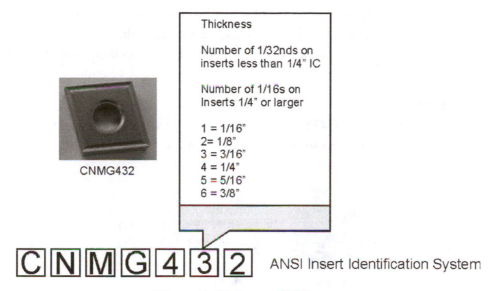

Thickness

Number of 1/32nds on inserts less than 1/4" IC

Number of 1/16s on Inserts 1/4" or larger

1 = 1/16"
2 = 1/8"
3 = 3/16"
4 = 1/4"
5 = 5/16"
6 = 3/8"

CNMG432

C N M G 4 3 2   ANSI Insert Identification System

Figure 2-32. Insert Thickness

Insert Corner Geometry (Tool Nose Radius)

The seventh position on the ANSI Insert identification system is the size of the corner or nose radius on the insert (see Figure 2-33). In our example we have a 2 in the 7th position. This would mean that this insert would have a 1/32 radius.

C N M G 4 3 2   ANSI Insert Identification System

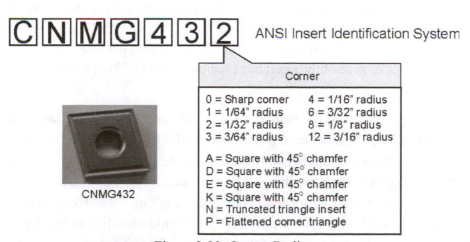

Corner

0 = Sharp corner	4 = 1/16" radius
1 = 1/64" radius	6 = 3/32" radius
2 = 1/32" radius	8 = 1/8" radius
3 = 3/64" radius	12 = 3/16" radius

A = Square with 45° chamfer
D = Square with 45° chamfer
E = Square with 45° chamfer
K = Square with 45° chamfer
N = Truncated triangle insert
P = Flattened corner triangle

CNMG432

Figure 2-33. Corner Radius

The number in the seventh position of the ANSI Insert identification system chart represents the number of sixty-fourths of nose radius. Although selecting the proper grade of insert is probably the most important, other factors such as nose radius are very important when selecting the proper tool for the application.

The nose radius of the tool directly affects tool strength and surface finish, as well as cutting speeds and feeds. The larger the nose radius: the stronger the tool. If the tool radius is too small, the sharp point will make the surface finish unacceptable, and the life of the tool will be shortened. Larger nose radii will give a better finish and longer tool life and will allow for higher feed rates. If the tool nose is too large, it can cause chatter. It is usually best to select an insert with a tool-nose radius as large as the machining operation will allow.

Insert Identification

Let's take one more look at our CNMG432 insert. What does this insert look like and what are the possible uses for this insert? Figure 2-34 shows a CNMG432 insert.

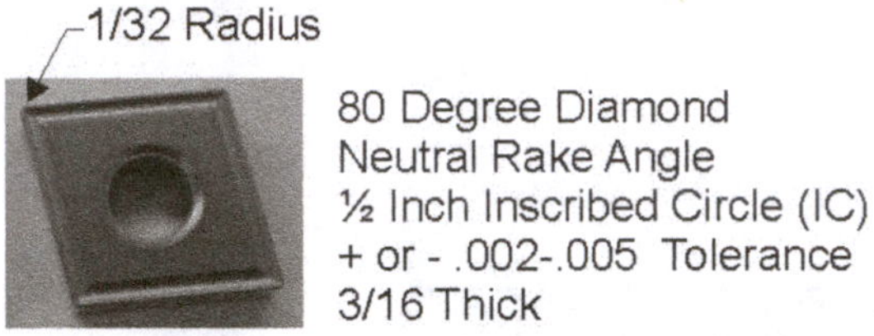

1/32 Radius

80 Degree Diamond
Neutral Rake Angle
½ Inch Inscribed Circle (IC)
+ or - .002-.005 Tolerance
3/16 Thick

Figure 2-34. A CMNG432 Insert.

The insert shape is designated as a C. This means that the shape of the insert is an 80 degree diamond. The clearance or relief angle designation is an N. This specifies that the insert has a zero or neutral relief angle and must be used with a negative relief holder. The third letter in the designation indicates the size tolerance of the insert. The M guarantees that the inserts repeatability on the inscribed circle is +/- .002-.005 of an inch and the thickness repeatability is +/- .005 of an inch. The G in the fourth position of the chart signifies that the insert has a molded chip breaker on both sides of the insert and that this insert would be held in a holder which has a lock pin. A 4 in the fifth position of the chart states that the insert has an inscribed circle size of 4/8 or one-half of an inch. The thickness of the insert 3/16 of an inch as indicated by the 3 in the sixth position of the chart. The last identifying number in the CNMG432 insert is a 2. This indicates the edge or nose of the insert has a radius of 2/64 (1/32) of an inch.

Possible Usage

The 80 degree diamond shape and the neutral relief of this insert would indicate that it will be used primarily for roughing. In most cases, finish stock of 0.02 to 0.03 is left on the part after roughing, so the size tolerance and repeatability wouldn't be an issue. A molded chip breaker would help with chip control when roughing. An inscribed circle of one-half inch and an insert thickness of 3/8 of an inch would allow us to take large roughing cuts. Keep in mind that the rigidity of the setup and the horsepower of the machine would also dictate the depth cut capabilities of this insert. The nose radius of this insert is 1/32 (2/64s) of an inch. This size nose radius is certainly capable of handling roughing feed rates and heavy depths of cuts.

Insert Selection

Now that we have covered some aspects of carbide tool selection, let's look at the questions that we need to answer when selecting the proper insert grade and style. One of the first considerations is the material to be machined.

Machinability of Metals

Machinability describes the ease or difficulty with which a metal can be cut. Machining involves removing metal at the highest possible rate and at the lowest cost per piece. Different materials' structures pose different problems for the machinist. Materials that are easy to machine have high machinability ratings and therefore cost less to machine. Materials that are difficult to machine have lower machinability ratings and cost more to machine.

The machinability of a material is directly related to the material's hardness. A number of tests measure the hardness of a material, but the most common test for machinability is the Brinell test. Brinell hardness, or BHN, is stated as a number: the higher the BHN number, the harder the material. Hardness, although a major factor affecting machinability, is not the only factor that determines machinability.

Steels

Steels are classified based on their carbon content and their alloying elements. Plain carbon steels have only one alloy, carbon, mixed with iron. Carbon has a direct effect on steel's hardness. Plain carbon steel's machinability is directly related to its carbon content. Alloy steels, on the other hand, have carbon and other alloying elements mixed with iron. These alloying elements can give steel the characteristic of not only being hard but also tough. The major concern with machining alloy steels is their tendency to work harden, a phenomenon that occurs when too much heat from the cutting process is developed in the steel. The heat changes the properties of the steel, making it harder and difficult to machine. Great care must be taken when machining some alloy steels.

Plain carbon steel is divided into three categories: low carbon, medium carbon, and high carbon. Low-carbon steels have a carbon content of 0.10 percent to 0.30 percent and are relatively easy to machine. Medium-carbon steels have a carbon content of 0.30 percent to 0.50 percent. Medium-carbon steels are relatively easy to machine, but because of the higher carbon content, they have a lower cutting speed than that of low-carbon steel. High-carbon steels have a carbon content of 0.50 percent to 1.8 percent. When the carbon content exceeds 1.0 percent, high-carbon steel becomes quite difficult to machine.

Stainless Steel

Stainless steels have carbon, chromium, and nickel as alloys. Stainless steels are a very tough, shock-resistant material and are difficult to machine. Work hardening can be a problem when machining stainless steels. To avoid work hardening, use lower speeds and increased feed rates. Chip control is sometimes a problem when machining stainless because of its toughness and the chips' unwillingness to break.

Cast Iron

Cast iron is a broad classification for gray, malleable, nodular, and chilled-white cast iron. This grouping is in order of its machinability. Gray cast iron is relatively easy to machine, while chilled-white cast iron is sometimes un-machinable. Cast iron does not produce a continuous chip because of its brittleness.

The machinability of any material can be affected by factors such as heat treatment. Heat treating can be used to harden or soften a material. The condition of the material at the time of machining should be taken into consideration when deciding a material's machinability.

Tool holder Style and Identification

Carbide manufacturers and the American Standards Association have created a tool holder identification system for indexable carbide tool holders (see Figure 2-35). Because a huge variety of holders are available, we cannot include all of the holder types in this text.

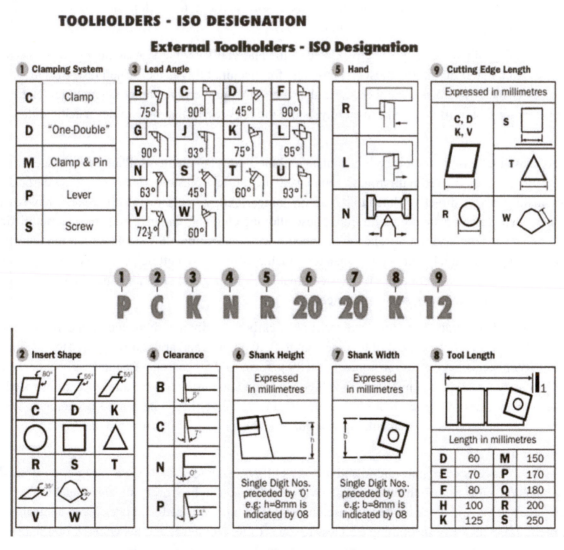

FIGURE 2–35. ISO Tool holder identification system.

Qualified Tooling

Tools that are used in CNC machines are machined to a high level of accuracy. Qualified tools are typically guaranteed to be within .003 of an inch.

The accuracy of the cutting tip is referenced to specific points or datums located on the holders. The higher level of accuracy enables the operator to change inserts without having to remeasure the tools. Figure 2-26 shows an example of where the measurements are qualified from.

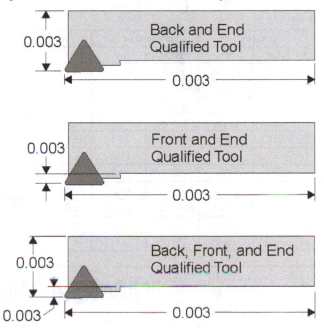

FIGURE 2–36. Qualification of tool holders.

Tool Insert and Tool holder Selection Practice

Figure 2–37 shows a typical lathe part. An appropriate insert grade could be chosen to machine the part using the ISO grade designation chart shown in Figure 2–18, the insert identification system shown in Figure 2–38 and 2-39, and an appropriate tool holder from Figure 2-35.

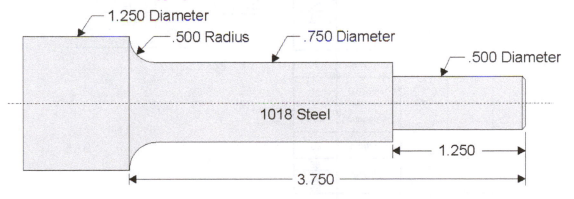

Figure 2-37. Lathe part.

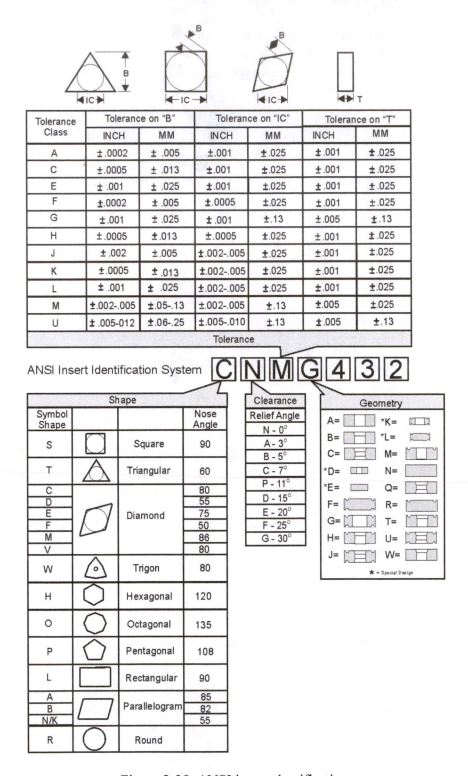

Figure 2-38. ANSI insert classification.

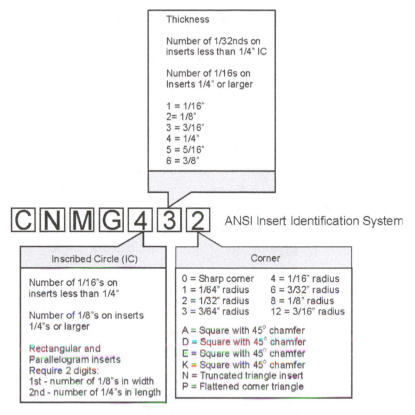

Figure 2-39. ANSI insert classification.

- What type of material is being cut? Would you use a cast iron or a steel grade? Answer: The material used for the part is 1018 cold rolled steel, so a steel cutting grade would be appropriate (see Figure 2-18).

- How hard is the material? How does this affect the grade? Answer: The material is a low-carbon alloy steel of only 200 Brinell hardness. A moderate hardness grade would be a good choice.

- What is the condition of the material? Does the surface show evidence of scale or hard spots? How does this affect the selection of the grade, insert shape, rake angle, and nose radius? Answer: The material is cold rolled steel, which has little or no scale. Again, a general-purpose insert of moderate hardness and strength would be applicable.

- What shape insert do we need to perform this job? Answer: For roughing this part, we would like to use a larger angled insert such as an 80-degree diamond. The finish tool needs to have a little smaller angle to cut the radius. A 55-degree diamond or triangular insert would be a good choice for finish cuts.

- How rigid is the machining setup? How does this affect the rake angles and nose radius? Answer: As you can see from the part print, the part has a small turned diameter on the end. This small diameter may tend to deflect and chatter. A positive rake insert with a 1/32 or 1/16 tool-nose radius would probably be the best choice.

- What are the surface finish requirements of the part? How does this affect the nose radius? Answer: The surface finish requirement of the part is 125. A 125 finish is a standard machine finish that can be held using a 1/32 or 1/16 tool- nose radius. Slowing down the feed rate will also help to acquire the 125 finish requirement.

Chip Control

Chip control refers to the ability to control the chip formation. Chip control is important to operator safety, tool life, and chip handling. CNC machines typically have chip conveyers to automatically deposit chips into recycling containers. If chips are allowed to become long and stringy, they will clog up the chip conveyers. Straight, stringy chips will also wrap around the tool, the workpiece, and the work-holding device, which can cause tool breakage and an especially dangerous situation for machine operators. Soft, gummy, and tough materials can wrap around spinning tools and workpieces. The chips begin whipping around, sending sharp, hot chips in every direction. For this reason, there has been considerable research in the area of chip control.

Molded Chip Breakers

Molded chip breakers use a molded groove to change the direction of the chip. Molded chip breakers are available in many different configurations (see Figure 2–40). Some molded chip breakers are designed for certain materials, and others are designed for different feed rates and depths of cut.

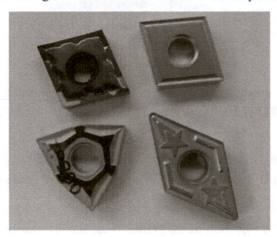

FIGURE 2–40. Indexable inserts with molded chip breakers.

The chip breaker is designed to redirect the flow of the chip, causing it to curl into a figure 6 or figure 9 (see Figure 2–41). A chip of this configuration is said to be the perfect chip for steel cutting. When cutting cast iron, chip control is not a problem because iron is brittle and does not flow away from the cutting edge the way steel does.

Factors Affecting Chip Formation

Chip breakers have greatly increased our ability to control chips. There are however factors that need to be addressed no matter which type of chip breaker is used. The three factors that affect chip control the greatest are feed rate, cutting speed, and tool shape. One of the quickest ways to eliminate long, stringy chips is to increase the feed rate: a thicker chip will curl and break easier than thin chips.

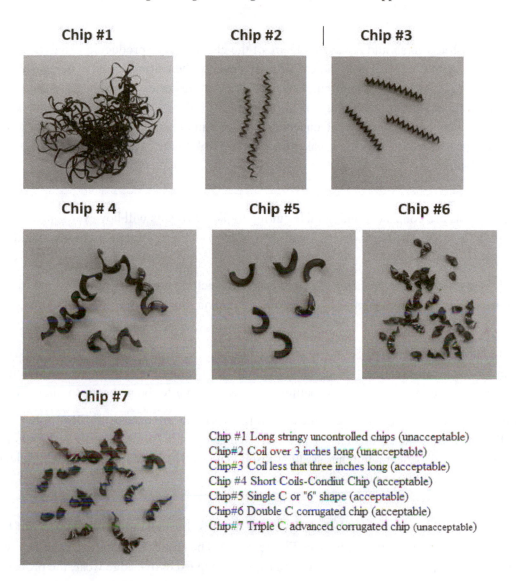

Chip #1

Chip #2

Chip #3

Chip # 4

Chip #5

Chip #6

Chip #7

Chip #1 Long stringy uncontrolled chips (unacceptable)
Chip#2 Coil over 3 inches long (unacceptable)
Chip#3 Coil less that three inches long (acceptable)
Chip #4 Short Coils-Condiut Chip (acceptable)
Chip#5 Single C or "6" shape (acceptable)
Chip#6 Double C corrugated chip (acceptable)
Chip#7 Triple C advanced corrugated chip (unacceptable)

FIGURE 2–41. Different chip configurations.

Decreasing the side-cutting edge angle or lead angle will also create a thicker chip. Sometimes increasing the speed will help the chip flow and curl easier. Long stringy chips are not the only chips that can cause problems in machining. Conduit chips are chips that are almost ready to break (see Figure 2–41). A conduit chip is a long curly chip, which is common when machining soft ductile materials. To remedy the problem, try increasing the feed rate.

Corrugated chips (see Figure 2–41) have a very tight curl and represent the opposite problem from stringy chips. These types of chips are being bent too much and are usually caused by excessive feed rates. Corrugated chips do not pose a chip control problem, but they are a sign of improper cutting action and should be dealt with immediately. If slowing the feed rate doesn't change the chip formation, use a narrower chip breaker to allow the chip to make a wider curl.

Chip Color

As a machine operator, you should always be aware of the chips you are producing. Analyze the chip's shape and color. A deep blue steel chip indicates that the heat of the cutting action is being drawn away from the workpiece, as it should. A dark purple or black chip indicates excessive heat. In this case, reduce the cutting speed and any other machining conditions until the color of the chip is acceptable. Chips should always be clean and smooth on the underside, not torn and ragged. Proper chip formation is a balancing act of speeds, feeds, and chip breaker formation. Look to the chip for the information you need to balance these factors.

Troubleshooting

Carbide cutting tools are consistent and durable cutting tools. Problems will however sometimes result when using carbide tools. The CNC operator will then need to change cutting conditions to address the problems.

The first step is to diagnose the problem. Possible problems include premature failure of the insert, edge wear, crater wear, edge buildup, depth of cut notching, chipping, thermal cracking, or thermal deformation.

Catastrophic Breakage

Premature failure or insert breakage is a problem that will be apparent even to the least experienced machine operator. If the insert breaks and continues to break even after being changed, there is a problem.

One possible cause of tool breakage is that the operating conditions are excessive. Slow down the speed and especially the feed. If the grade that you have selected is too hard (brittle) for the material or the condition of the material, select a tougher grade of insert. The lead angle may be too small. Select a tool holder that lends more support to the tool tip.

Edge Wear

Edge wear is more difficult to diagnose. Excessive edge wear is the unnatural wearing away of the insert along the side or flank of the cutting edge (see Figure 2–42). The ability to recognize excessive edge wear comes with experience. If you believe that you are experiencing excessive edge wear, the probable cause is friction. Excessive friction causes heat to build along the cutting edge, which causes the binders to fail. One possible cause is that the lead angle is too great. Choose a holder that reduces the lead angle. Check the tool height. A crash or bump of the tool turret may be causing the tool to be too high. Another possible cause may be that the feed rate is too low. Increasing the feed rate will cause the chips to concentrate away from the cutting edge. Finally, it could be a grade selection problem.

Crater Wear

Crater wear occurs when the binder is being replaced by the material you are cutting. When you are cutting steel, the constant passing of the chip over the insert causes the cobalt binder to be carried away by the chip, leaving the steel to act as the binder. The steel, not being a very good binder material, quickly wears away, leaving a crater (see Figure 2–43). Cratering is usually a grade selection problem or an extreme heat problem, caused by cutting conditions that are too high. To minimize cratering, reduce the speed and feed, use a harder grade of carbide, or use a coated carbide insert.

FIGURE 2–42. Edge Wear.

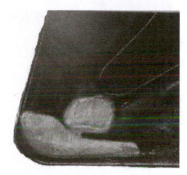

FIGURE 2–43. Crater Wear.

Edge Buildup

Edge buildup or adhesion occurs when metal deposits build up on the cutting edge (see Figure 2–44). Iron actually combines with the binder in the carbide substrate. Edge buildup occurs when the cutting conditions are too slow. Carbide cuts best at high temperatures and will rapidly wear if these temperatures are not reached. Machine operators can reduce edge buildup by increasing the speed and feed.

FIGURE 2–44. Edge Buildup.

Depth-Of-Cut Notching

Depth-of-cut notching is an unnatural chipping away of the insert right at the depth of cut line (see Figure 2–45). Depth-of-cut notching is usually a grade selection problem. If you are using an uncoated insert, consider changing to a coated insert. If a coated insert is not available, try honing the edge of the insert. Honing should only be done on uncoated inserts. Honing is done at a 45-degree angle to the cutting edge. Proper honing just breaks the sharp edge of the insert. Depth-of-cut notching may also be solved by lowering the feed rate and/or by reducing the lead angle.

FIGURE 2–45. Depth of cut notching.

Chipping

Chipping is a common insert problem. Chipping occurs along the cutting edge and is sometimes mistaken for edge wear (see Figure 2–46). The major causes of insert chipping are lack of rigidity, too hard of an insert grade, and low operating conditions. Carbide is very brittle and works best when it is well supported.

By decreasing the overhang of the tool and supporting the work better, the operator can eliminate many carbide cutting tool problems. If rigidity is not the problem, use a softer or tougher grade of insert. When making roughing cuts through hard spots or sand inclusions, use a tougher, not harder, grade of carbide. If the operating conditions are too low, abnormal pressures may build up, causing chipping. Increasing the cutting speed will sometimes eliminate chipping.

FIGURE 2–46. Chipping.

Thermal Cracking and Thermal Deformation

Two heat problems are commonly associated with carbide cutting tools: thermal cracking and thermal deformation (see Figure 2–47). Thermal cracking will show up as small surface cracks along the cutting edge and tip of the insert. Cracking is caused by sudden changes in temperature. Thermal cracking can occur if coolant is being applied to the insert instead of in front of the insert. If coolant is applied in the middle of the cut, thermal cracking may occur.

FIGURE 2–47. Thermal Cracking.

The other heat-related problem associated with carbide cutting tools is thermal deformation. Thermal deformation is a melting away of the tool tip and is caused by operating conditions being too high (see Figure 2–48). The excessive heat breaks down the binder materials in the carbide insert. There are two possible solutions to thermal deformation: reduce the cutting conditions or switch to a more heat-resistant grade of carbide.

FIGURE 2–48. Thermal Deformation.

When trying to diagnose problems with carbide cutting tools, remember that troubleshooting is not a shot in the dark and should be done systematically. Troubleshooting must be a methodical procedure. The first step is to determine the problem. The second step is to arrive at all of the possible solutions. The third step is to examine each of the possible causes, changing only one condition at a time.

Use Figure 2–49 to help diagnose your carbide cutting tool problems.

Problem	Remedy
Tool life is too short due to excessive wear	1. Change to a harder, more wear resistant grade. 2. Reduce the cutting speed. 3. Reduce the feed. 4. Increase the lead angle. 5. Increase the relief angle.
Excessive Cratering	1. Use a harder, more wear resistant grade. 2. Reduce the cutting speed. 3. Reduce the feed.
Cutting edge chipping	1. Increase the cutting speed. 2. Hone the cutting edge. 3. Change to a tougher grade. 4. Use a negative rake insert. 5. Increase the lead angle. 6. Reduce the feed.
Deformation of the cutting edge	1. Reduce the cutting speed. 2. Change to a grade with a higher red-hardness. 3. Reduce the feed.
Poor surface finish	1. Increase the cutting speed. 2. Increase the nose radius. 3. Reduce the feed. 4. Use positive rake inserts.

FIGURE 2–49. Troubleshooting chart. Find your machining problem on the left, and the right column will list potential cures for the problem.

Figures 2-50 and 2-51 show the ANSI insert classification system. They will be helpful for the chapter questions.

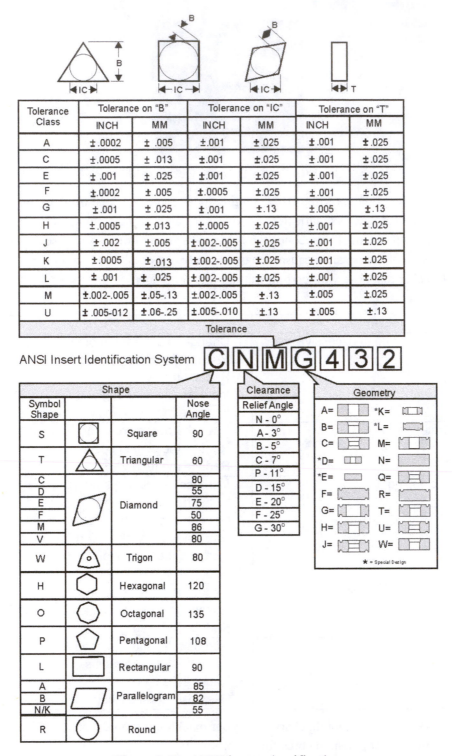

Tolerance Class	Tolerance on "B"		Tolerance on "IC"		Tolerance on "T"	
	INCH	MM	INCH	MM	INCH	MM
A	±.0002	±.005	±.001	±.025	±.001	±.025
C	±.0005	±.013	±.001	±.025	±.001	±.025
E	±.001	±.025	±.001	±.025	±.001	±.025
F	±.0002	±.005	±.0005	±.025	±.001	±.025
G	±.001	±.025	±.001	±.13	±.005	±.13
H	±.0005	±.013	±.0005	±.025	±.001	±.025
J	±.002	±.005	±.002-.005	±.025	±.001	±.025
K	±.0005	±.013	±.002-.005	±.025	±.001	±.025
L	±.001	±.025	±.002-.005	±.025	±.001	±.025
M	±.002-.005	±.05-.13	±.002-.005	±.13	±.005	±.025
U	±.005-012	±.06-.25	±.005-.010	±.13	±.005	±.13
Tolerance						

ANSI Insert Identification System C N M G 4 3 2

Shape

Symbol Shape			Nose Angle
S		Square	90
T		Triangular	60
C		Diamond	80
D			55
E			75
F			50
M			86
V			80
W		Trigon	80
H		Hexagonal	120
O		Octagonal	135
P		Pentagonal	108
L		Rectangular	90
A		Parallelogram	85
B			82
N/K			55
R		Round	

Clearance

Relief Angle
N - 0°
A - 3°
B - 5°
C - 7°
P - 11°
D - 15°
E - 20°
F - 25°
G - 30°

Geometry

A= *K=
B= *L=
C= M=
*D= N=
*E= Q=
F= R=
G= T=
H= U=
J= W=

★ = Special Design

Figure 2-50. ANSI insert classification.

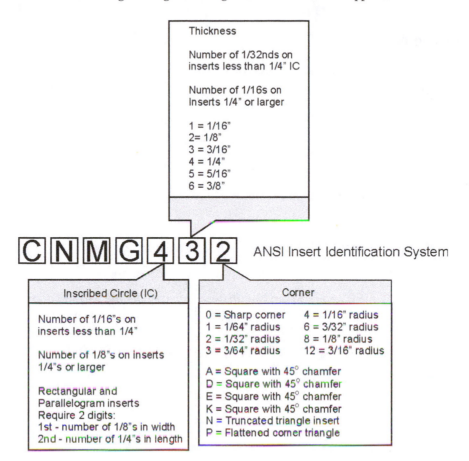

Figure 2-51. ANSI insert classification.

Chapter Questions

1. Explain what is meant by the term cutting speed.

2. How is cutting speed derived?

3. What is the formula for calculating RPM?

4. What is the formula for feed rate in inches per minute?

5. State the two main characteristics of carbide.

6. What is meant by *insert grade?*

7. What is a coated carbide?

8. Name three types of coating that are applied to carbide inserts.

9. Name five different insert shapes in order of increasing strength.

10. What is one of the most common binding materials that holds the carbide particles together?

11. State the purpose of tool-nose radius.

12. Name two factors to consider when selecting the proper shape of carbide insert.

13. What is meant by *inscribed circle?*

14. Describe the three types of back rake angles.

15. What is lead angle?

16. What are qualified tool holders?

17. Describe the characteristics of a TPG 432 insert.

18. Describe the characteristics of a VNMG 332 insert.

19. Describe the characteristics of a CNMG 432 insert.

20. Describe the characteristics of a SPG 432 insert.

21. While face milling a previously drilled surface you notice that the inserts start chipping as soon as the face milling cutter enters the interrupted cut. What are some possible remedies for this type of tool failure?

22. After rough turning with a new insert for five minutes you notice that the insert is starting to spark and the part is getting very warm. What type of tool failure is this and what are some possible remedies for this type of tool failure?

23. While attempting to turn a piece of cast steel the tool tip keeps breaking off. What are some possible remedies for this type of tool failure?

24. While milling mild steel you notice that the chips are sticking to the inserts. What are some possible remedies for this machining situation?

25. While turning tool steel you notice that there is a groove appearing on the insert. What are some possible remedies for this type of tool failure?

Chapter 3

FUNDAMENTALS OF MACHINING CENTERS

INTRODUCTION

A CNC machining center is a CNC milling machine that is equipped with an automatic tool changer. The manual milling machine is a very versatile and productive machine tool, but when coupled with a computer control, it becomes the production center of the machine shop. Repetitive operations, such as drilling, tapping, and boring, are perfect applications for the machining center.

OBJECTIVES

Upon completion of this chapter, the reader will be able to:
- Describe the purpose and function of the machining center.
- Identify the major components of the machining center.
- Identify the axes and directions of motion on a typical machining center.
- Describe work-holding devices used on machining centers.
- Differentiate between climb and conventional milling.
- Calculate speeds and feeds for milling.
- Describe the different methods of manually moving the machine axes.
- Define terms such as "MDI" and "conversational."

Types of Machining Centers

Machining centers are identified by the orientation of the spindle. Machining centers are either horizontal or vertical spindle machines.

Horizontal Machining Centers
Horizontal machining centers are typically the workhorses in the shop and are patterned after horizontal boring mills (see Figures 3–1). The horizontal configuration of the spindle lends itself to heavy cuts on large workpieces. Horizontal machining centers often have twin tables known as pallets. While one pallet is within the machining envelope, the other pallet is outside of the machining area, allowing the next piece to be setup on the pallet while the part on the other pallet is being machined.

FIGURE 3–1. The horizontal spindle machining center is best suited for heavy machining operations on large workpieces. Courtesy Haas Automation Inc.

Vertical Machining Center
The vertical machining center is probably the most versatile and common CNC machine found in the machine shop (see Figures 3-2). The vertical configuration of the spindle lends itself to quick, easy workpiece setups. There are a variety of types and sizes of vertical machining centers. The type of machining to be done and the size of the work will determine which machine is best for your application.

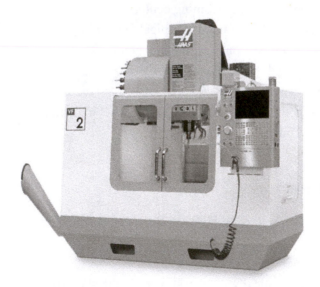

FIGURE 3–2. Vertical spindle machining centers are very versatile machines. The quick setup of these machines makes them very popular in the machine shop. Courtesy Haas Automation Inc.

Parts of the Machining Center
Figure 3–3 shows the main components of a vertical machining center.

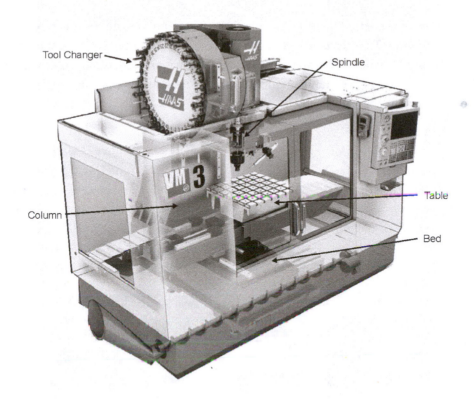

FIGURE 3–3. The main components of the vertical machining center. Courtesy Haas Automation Inc.

Column

The column is the backbone of the machine. The column is typically mounted to the saddle and provides one of the axes or directions of travel. The rigid construction of the column will keep the machine from twisting during machining.

Bed

The bed is one of the more integral parts of the machining center. The bed is typically produced from high-quality cast iron, which absorbs the vibration of the machining operation. Hardened and ground slideways are mounted to the bed to provide alignment and support for the machine axes.

Table

The table is mounted on the bed, and the work or a work-holding device is mounted to the table. The table has T-shaped slots milled in it for mounting the work or work-holding device.

Spindle

The spindle holds the cutting tool and is programmable in revolutions per minute.

Tool Changers

Tool changers are an automatic storage and retrieval system for the cutting tools. An automatic tool changer makes the CNC milling machine a machining center.

Carousel-Type Tool Changers

Carousel-type tool changers are spindle-direct tool changers, meaning they do not use auxiliary arms to change tools (see Figure 3–4). The carousel can be mounted on the back or side of the machine. Carousel tool changers are typically found on vertical machining centers. When a tool change is commanded, the machine moves to the tool change position and puts the current tool away. The carousel then rotates to the position of the new tool and loads it.

Arm-Type Tool Changers

The tool change arm rotates between the machine spindle and the tool magazine. After getting the tool change command, the tool that is in the spindle will come to a fixed position known as the "tool change position". The Automatic Tool Change (ATC) arm will rotate and will pick up the tool using a gripper. The arm has two grippers, one on each end of the arm. Each gripper can rotate through 90°, to deliver tools to the spindle. One end of the arm will pick up the old tool from spindle and the other end will pick up the new tool from the tool magazine. The arm then rotates and places the old tool back in the tool magazine.

FIGURE 3–4. Carousel-type tool changing systems are usually found on vertical machining centers. The spindle positions itself over the tool and then moves down and clamps the tool in the spindle. Photo on the right, Courtesy Haas Automation Inc.

Tool carousels typically have bidirectional capabilities, which allows for quicker tool changes. Tool-changing cycle time is very important to a machining center's productivity. Tool-changing time is nonproductive time. No machining occurs while a tool is being changed.

Axes of Motion

The linear axes or directions of travel of the machining center are defined by the letters X, Y, and Z (see Figure 3–5). Axis designation letters appear with positive or negative signs for direction of travel. The Z axis always lies in the same direction as the spindle. A negative Z (–Z) movement always moves the cutting tool closer to the work. The X axis is usually the axis with the greatest amount of travel. On a vertical machining center, the X axis would be the left/right travel of the table. The Y axis on the vertical machining center would be the travel toward and away from the operator. On the horizontal machining center, the Y axis would be the up and down travel of the spindle head. It may be helpful to think of the motion in terms of tool position. If the table is moved to the left, the tool is positioned more in the + X direction. If the table is moved toward the front, the tool is positioned more in the +Y direction.

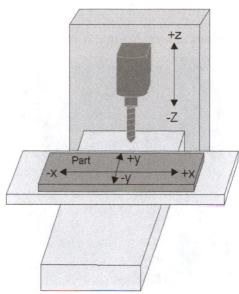

FIGURE 3–5. The three linear axes of the vertical machining center are identified as X, Y, and Z. The Z axis always lies in the same plane as the spindle.

Rotational Axes

Some CNC machining centers are capable of rotary cutting motions. A horizontal machining center that is equipped with a rotary table is capable of four axes of motion. The fourth axis is known as the C axis (see Figure 3–6). The C axis is a rotational axis about the Z axis. Machining centers that are equipped with a rotary table and tilting, contouring spindle are said to have five axes: three linear and two rotary. Four- and five-axis machines are used to machine parts with complex surfaces such as mold cavities or rotary turbines.

FIGURE 3–6. The rotary table adds a fourth axis of travel to the machining center. The rotary table can be used to do helical-type machining or it can be used to reposition the workpiece. Courtesy Haas Automation Inc.

Machine Control Features

Machining center controls come in all styles and levels of complexity, but they all have many of the same features. If you have a good understanding of one machine, the next control will be that much easier. Control functions are divided into two distinct areas: manual controls and program controls.

Manual Control
Manual control features are those buttons or switches that control machine movement (see Figure 3–7).

FIGURE 3–7. The manual machine controls are located on the left-side of this control.

Emergency Stop Button
The emergency stop button is the most important component on the machine control. This button has saved more than one operator from disaster by shutting down all machine movement.

This big red button with the word "Reset" or "E-stop" on the front should be used when it is evident that a collision or tool breakage is going to occur. Emergency stop buttons are located in more than one area on the machine tool and should be located prior to doing any machine operations.

Moving the Axes of the Machine
Manual movement of the machine axes is done in a number of different ways. Most controls are equipped with a pulse-generating hand wheel (see Figure 3–8).

FIGURE 3–8. Some controls have a hand wheel for each axis, while other controls have one hand wheel and a switch to select which axis the operator will move.

The hand wheel has an axis selection switch that allows the operator to choose which axis he or she wants to move. The handle sends a signal or electronic pulse to the motors, which move the table or the spindle head. If the handle of the Y axis is turned in the negative direction, the tool moves toward the operator. If the Z axis hand wheel is moved in a negative direction, the tool moves toward the table.

Some machines are equipped with jog buttons (see Figure 3–9). When the jog button for a certain axis is pressed the axis moves. The distance or speed at which the machine moves is selected by the operator prior to the move.

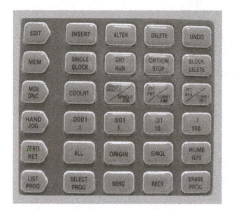 

FIGURE 3–9. Jog buttons are used for machine axis movement. The amount of movement per push of the jog button is controlled by the mode selector switch.

Spindle Speed and Feed Rate Override Switches

Spindle speed and feed rate overrides are used to speed up or slow down the feeds and speeds of the machine while machining (see Figure 3-10). The override controls are typically used by the operator to adjust to changes in cutting conditions, such as hard spots in the material. Feed rates can typically be adjusted from 0 to 150 percent of the programmed feed rate. Spindle speeds can typically be adjusted from 0 to 200 percent of the programmed spindle speed.

FIGURE 3–10. The spindle speed and feed rate override switches give the operator a greater amount of control over the machine.

Cycle Start/Feed Hold Buttons

The two most commonly used buttons on the control are the cycle start and feed hold buttons (see Figure 3–11). The cycle start button is used to start execution of the program. The feed hold will stop execution of the program without stopping the spindle or any other miscellaneous functions. By pushing the cycle start, the operator can restart the execution of the program.

FIGURE 3–11. The cycle start and feed hold buttons are located adjacent to each other. These buttons will typically light up when activated. The feed hold button will only stop the axis from traveling; it will not stop the spindle or any other miscellaneous function.

Home or Zero Return

The home or zero return button, when selected, will return all of the axes of the machine to the home position. Every CNC machine has an assigned stationary position known as home or reference zero. Home position is usually defined as a position at the extreme travel limits of the three main axes. The zero or home position is set by using switches and encoders. When the machine is at home position, the sensors are triggered and the axes indicator lights illuminate. The home return button is used when the operator wants to load or unload parts and when he or she wants to start the program from the home position. Before the operator shuts the machine down, the axes of the machine should be brought back to the home position. This will ensure quick and easy home positioning of the machine upon startup.

Workpiece Coordinate Setting

The workpiece coordinate or program zero is the point or position from which all of the programmed coordinates are established. For example, if the programmer looks at the part print and notices that all of the dimensions come from the center of the part, these datums are then used to establish the program zero or workpiece coordinate (see Figure 3-12).

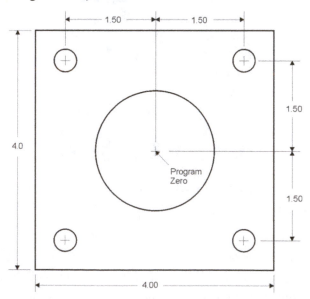

FIGURE 3–12. The workpiece coordinate system can be set from any datum feature. Pick the feature that would allow the programmer to do the fewest calculations.

The part origin is the X0, Y0, Z0 location of the part in the rectangular or Cartesian coordinate system. In absolute programming, all of the tool movements would be programmed with respect to this point. If all of the dimensions were located from the center of the bored hole, then that point would become the program zero.

During part setup, the X and Y zero position of the part has to be located. Using the hand wheels or other manual positioning devices and an edge finder or probe, the setup person locates the point at which the center of the spindle and the part origin are the same (see Figure 3-13).

The "home zero" is then entered as a G-code in the appropriate area of the program or in an offset table.

FIGURE 3–13. It is extremely important to accurately locate the workpiece zero point; otherwise, all of the machined features will be shifted out of location. An edge finder (left) or probe (right) is used to position the center of the spindle at the part zero location.

The setup person must then measure and enter the values of the tool lengths in the offset table for each tool being used in the program. Each tool used in the program has a different length.

The control must then be told to compensate for the difference in the lengths. The tool length offsets are typically the distance from Z at machine zero to the Z position of the part zero (see Figure 3–14). The tool length offsets are stored as an offset in the control (see Figure 3–15). This process may seem a little confusing, but once you have done it, it really is quite straightforward.

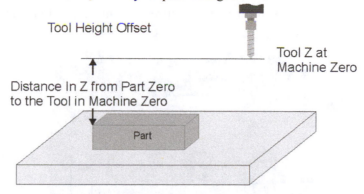

FIGURE 3–14. The operator touches the tool off on the Z0 point of the part. He or she then takes the Z axis distance from the Z0 point of the part and the machine home position. The distance is recorded in the tool length offset table. This must be done for every tool that is used in the program.

<< PROBING	TOOL OFFSET			TOOL INFO >>	
TOOL 1	COOLANT	H(LENGTH)		D(DIA)	
OFFSET	POSITION	GEOMETRY	WEAR	GEOMETRY	WEAR
1 SPINDLE	10	4.5680	0.	0.	0.
2	0	0.	0.	0.	0.
3	0	0.	0.	0.	0.
4	0	0.	0.	0.	0.
5	0	0.	0.	0.	0.
6	0	0.	0.	0.	0.
7	0	0.	0.	0.	0.
8	0	0.	0.	0.	0.
9	0	0.	0.	0.	0.

<< WORK PROBE		WORK ZERO OFFSET		WORK PROBE >>	
G CODE	X AXIS	Y AXIS	Z AXIS		
G52	0.	0.	0.		
G54	-12.5680	-8.4890	-23.1480		
G55	0.	0.	0.		
G56	0.	0.	0.		
G57	0.	0.	0.		
G58	0.	0.	0.		
G59	0.	0.	0.		
G154 P1	0.	0.	0.		
G154 P2	0.	0.	0.		
G154 P3	0.	0.	0.		

ENTER A VALUE

FIGURE 3–15. When the program calls for the tool length offset, the control accesses the register located in the tool offset area. The control uses this to compensate for the tool length.

Single Block Operation

The single block option on the control is used by the operator to advance through the program one block or line at a time. When the single block switch is on, the operator presses the button each time he or she wants to execute a program block.

When the operator wants the program to run automatically, he or she can turn the single block off and press cycle start, and the program will run through without stopping. The purpose of the single block switch is to allow the operator to watch each operation of the program carefully. It is typically used the first time a program is run.

Manual Data Input

Manual data input or MDI is a means of inputting commands and data. MDI can be used to enter a simple command, such as starting the spindle, or to enter an entire program and is done through the alphanumeric keyboard located on the control (see Figure 3–16).

FIGURE 3–16. The alphanumeric keyboard lets the operator edit or enter a program at the machine control.

Program Editing

After the part program is loaded into the control, it may need some modification. The need for the change usually shows up on the shop floor. The operator or programmer can make changes using the program edit mode. The programmer uses the display screen to find the program errors and the keyboard to correct the errors.

Display

The display shows information such as the written program or part graphics. The program may be too long to fit on the screen, so it is separated into pages. The page or cursor button allows the operator to move through consecutive parts of the program. Graphics can also be displayed on the screen if your machine has graphics capabilities. Graphics are a representation of the part and the tool path, which would be generated by the program. Graphics are a safe way to test part programs. The simulation of the tool path will allow you to see program errors quickly, such as misplaced decimal points or missing minus signs. Graphics are used extensively in conversational programming.

Conversational Programming

Conversational programming is a built-in feature that allows the programmer to respond to a set of questions that are displayed on the graphics screen. The questions guide the programmer through each phase of machining. First, the operator might input the material to be machined. This information will be used by the control to calculate speeds and feeds. The operator can override these. Next, the operator would choose the operations to be performed and input the geometry. Operations such as pocket milling, grooving, drilling, and tapping can be performed.

With each response to a question, more questions are presented until the operation is complete. By answering the questions, the programmer is filling in variables in a canned or preprogrammed cycle. This type of programming is quicker than methods using word address programming. There is no standard conversational part programming language, and each system can be quite different. Once the programmer completes the conversational program, many CNC controls convert the conversational language into standard Electrical Industries Association/International Standards Organization (EIA/ISO) word address language. That is why this text will concentrate heavily on EIA/ISO word address programming language.

Diagnostics

The diagnostics mode consists of routines that detect errors in the machine system. An error number and message will be displayed on the screen. If any error is found in the CNC operation or servo system, the error message will prompt the operator or service technician to the cause of the problem.

This chapter has given you some basic insights into the machining center. Use this information as a place to start; take some time with an operator or instructor and get to know the operation of the control on your machining center.

Safety

Remember that no one has ever thought that they were going to be injured. But it happens! It can happen in a split second when you are least expecting it. An injury can affect you for the rest of your life. You must be safety minded at all times. Please get to know your machine before operating any part of the machine. Keep these safety precautions in mind.

Chapter Questions

1. Describe the function of the tool changer.
2. What are the three major axes associated with the machining center?
3. Which axis always lies in the same plane as the spindle?
4. What purpose does a pallet-changing system perform?
5. What piece of equipment gives a machining center a 4th axis of motion?
6. Describe two methods of manually moving the axes of the machining center.
7. Describe the workpiece coordinate or zero point.

Chapter 4

WORK HOLDING AND TOOLING FOR MACHINING CENTERS

INTRODUCTION

The proper choice and use of work holding and tooling are essential in making high-quality parts in an efficient and safe manner. This chapter will examine common methods of holding workpieces and the tooling that is commonly used in machining centers.

OBJECTIVES

Upon completion of this chapter, the reader will be able to:
- Describe various work-holding devices commonly used on machining centers.
- List common tools that are used on machining centers.
- Choose tooling for a machining operation.
- Differentiate between climb and conventional milling.
- Describe the advantages and disadvantages of climb milling and conventional milling.

Work Holding Devices

Work-holding techniques are very important in the setup and operation of a machining center. Before any machining can be done, the operator must make sure that the part or work-holding device is properly positioned and fastened to the table. Some setups may be as simple as placing the part in a vise, but some setups may take considerable ingenuity and time. Whatever the case, it is important to remember to make your setup as safe as possible.

Vises

The vise is the most common work-holding device (see Figure 4–1). The plain vise is used for holding work with parallel sides and is bolted directly to the table using the T-slots in the machine table. Air or hydraulically operated vises can be used in high-production operations to increase productivity.

FIGURE 4–1. The standard milling machine vise is used to hold relatively small parts that have a square or rectangular shape.

Angle Plates

Work that needs to be held at a 90-degree angle to the table can be held on an angle plate (see Figure 4–2). An angle plate is an L-shaped piece of cast iron or steel that has tapped holes or slots.

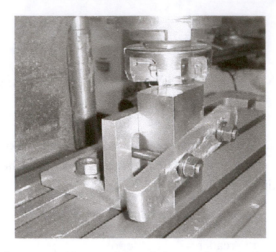

FIGURE 4–2. Angle plates come in a variety of sizes and are typically bolted directly to the machine table.

Direct Workpiece Mounting

Work that is too big or has an odd configuration can be bolted directly to the table (see Figure 4–3). This method of work holding takes the most ingenuity and expertise. There are a number of accessories that can be used to aid the setup person. A variety of clamp styles are commercially available for directly mounting workpieces to the machine table (see Figure 4–4). Figure 4–5 shows some clamping practices. Following are some tips for clamping work directly to the table:

1. Tables should be protected from abrasive materials, such as cast iron, by placing plastic or aluminum shims between the work and the table.

2. Clamps should be located on both sides of the workpiece if possible.

3. Clamps should always be located over supports to prevent distortion or breakage of parts.

4. Clamps and supports should be placed at the same height.

5. Screw jacks should be placed under parts for support to prevent vibration and distortion.

FIGURE 4–3. Parts that are clamped directly to the table are typically of an odd configuration, such as a weldment. Setting up these types of workpieces takes some ingenuity.

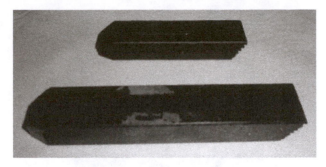

FIGURE 4–4. Strap clamps are used to fasten work to the machine table, fixture, or angle plate. Strap clamps are usually supported by step blocks. T-bolts should be placed as close to the workpiece as possible.

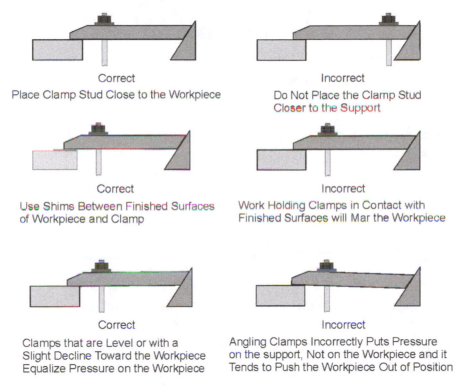

FIGURE 4–5. Study these acceptable clamping practices carefully.

Fixtures

Fixtures are tools that are designed and built to hold and accurately position a specific part. They are typically found in a production shop and can be built to hold one part or thousands of parts, depending on the application. The fixtured part is usually one that cannot be held in a vise (see Figure 4–6). Fixtures are used quite extensively in the machining industry. They should be kept simple to allow for quick loading and unloading of parts because loading time is unproductive time. Fixtures also need to be designed and built in a foolproof manner so that the part can be loaded in only one way. A well-designed fixture will lower the cost of producing parts.

FIGURE 4–6. Fixtures are used to hold and accurately position workpieces. Repeatability in locating the part is very important.

Tooling

Milling and drilling tools make up the majority of the types of tools used on the machining center. This section will discuss standard types of tools and tool holders used on milling machines. High speed steel tools and carbide tools will be covered in this section. Carbide tooling is crucial to the productive use of CNC machines.

High-Speed Steel Drills

The two basic types of high-speed drills are the twist drill and the spade drill. The high-speed steel twist drill is the most commonly used tool for producing holes.

Twist drills are great for rapidly producing holes that do not have to be very accurate in size or position. If the holes must be very accurate in size they are drilled to a smaller size and reamed, milled, or bored to size. If the position of the hole must be very accurate, the drilled hole must be milled or bored on location.

Twist drills are made with two or more flutes and come in a variety of styles (see Figure 4–7).

FIGURE 4–7. Twist drills are the most common hole-producing tools in the machine shop.

Twist drills have either a straight or tapered shank. Straight-shank drills are common up to 1/2 inch in diameter and are held in drill chucks.

Larger drills typically have a tapered shank with a tang on the end (see Figure 4–8). The tang keeps the tapered shank drill from slipping under the higher torque conditions associated with drilling large holes.

FIGURE 4–8. The tang on the end of the tapered shank drives the drill.

Center or Spotting Drills

When drilled holes need to be accurately located, it is advisable to center or spot drill the holes prior to drilling. This spotting or centering is achieved with center or spotting drills (see Figure 4–9). These drills are short, stubby, and rigid and do not flex or deflect, as longer drills have a tendency to. The spot drill produces a small start point that is accurately located. When the hole is drilled, the drill point will follow the starting hole that the spot drill made. This method can produce holes that are reasonably accurate in location.

FIGURE 4–9. Center and spotting drills come in a variety of sizes. The short, stubby design of the drills allows them to accurately locate holes.

Spade Drills

The spade drill has a flat blade with sharpened cutting edges (see Figure 4–10). The spade cutting tool is clamped in a holder and can be resharpened many times. Spade drills typically are used for drilling very large diameter holes. They can lower tooling costs because standard blade holders can hold a variety of sizes of blades. Designed to drill holes in one pass, spade drills require approximately 50 percent more horsepower than twist drills. Spade drilling also requires a rigid machine and setup.

FIGURE 4–10. Spade drills are a two-piece tool consisting of the blade and the holder. Spade blades and holders increase productivity because one holder can be used to drill many different diameter holes simply by changing blades.

Carbide Drills

Carbide-tipped twist drills have been around for many years. They are basically carbon steel drills with a piece of tungsten carbide brazed into them. They look similar to a spade drill but are usually made in smaller diameters. Solid carbide drills are just that, a solid carbide cutting tool in a twist configuration. Solid carbide drills are typically found in small diameters because of the cost of the carbide materials.

One of the newer innovations in carbide drilling technology is carbide insert drills (see Figure 4–11). These drills incorporate indexable or replaceable inserts and can remove metal four to ten times faster than a high-speed steel drill. Carbide insert drills require a rigid setup and a machine with substantial horsepower.

FIGURE 4–11. Carbide insert drills allow you to drill hard materials at feeds and speeds much higher than those of conventional drills. When the drill becomes dull, the carbide inserts can be indexed or replaced.

Auxiliary Hole-Producing Operations

Drilling may be the most common method of producing holes, but it is not the most accurate. In some cases, holes may need a very accurate size and/or finish. If an accurate-size hole is needed, reaming may be the quickest method.

A reamer is a cylindrical tool similar in appearance to the drill (see Figure 4–12). Reamers produce holes to t tight tolerance with a smooth finish. A slightly smaller hole must be drilled in the part before the hole can be reamed. The reamer follows the drilled hole, so inaccuracies in location cannot be corrected by reaming. If accurate location is needed as well as accurate size, boring may be necessary. Reamers are a quick way of producing accurately sized holes. Boring can produce holes of any size with a good finish and can locate them very accurately.

FIGURE 4–12. Reamers consistently produce accurately sized holes. The rule of thumbs for reamer feeds and speeds is that reamers should be run at half the speed and twice the feed of the same size drill.

Boring

Boring is done with an offset boring head and cutting tool. Figure 4–13 illustrates a boring head that utilizes a carbide insert. The offset boring head holds the tool and can be adjusted to cut any size hole within its range.

FIGURE 4–13. The offset boring head can be adjusted to cut any size hole within its size range.

Most boring heads are made to accommodate boring bars. Figure 4–14 shows a typical boring bar that could be used in a offset boring head. Boring accurately produces holes of any size and in the exact location.

FIGURE 4–14. Boring bars are tools used to do the cutting in a boring operation.

Boring, similar to reaming, can be done only on previously prepared holes. As a general rule, for best results, the boring tool should be as short and as large in diameter as possible. When using a high-speed steel tool, the diameter-to-length ratio should be no greater than 5-to-1. Example: If a 1-inch-diameter boring bar is used, no more than 5 inches should be sticking out. Carbide has a 3-to-1 recommended ratio. Reducing the ratio helps insure against chatter because as the tool overhang becomes shorter, the amount of force it takes to flex the boring bar increases.

Tapping

Tapping is the process of producing internal threads by using a tap. There are many different types of taps (see Figure 4–15). The most common type of tap used on the machining center is the spiral pointed gun tap, which is especially useful for tapping holes that go through the workpiece or holes with sufficient space for chips. Chip clearance is especially important when tapping. If chips clog the hole, a broken tap often results.

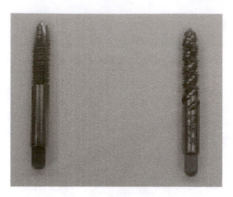

FIGURE 4–15. The two most common types of machine taps used on CNC machines are the gun tap and the spiral fluted tap. Gun taps push the chips ahead of the tap, so you must consider the amount of chip clearance that is available. The spiral flutes on a spiral-flute tap enable lubricant or coolant to reach the end of the tap.

There are two ways to tap on CNC machines. One way uses a special tapping head. The tapping head is spring loaded, and the lead of the tap provides the primary feed (see Figure 4–16). The secondary or programmed feed need only be approximate because the spring-loaded head allows the tap to float up and down at the lead rate of the thread. The lead of the tap is the distance that the thread travels in one revolution (see Figure 4–17). The lead of a thread and the pitch of the thread are the same for a regular-single lead thread.

FIGURE 4–16. The spring-loaded tapping head allows the tap to feed down at its own rate.

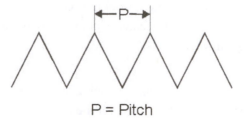

FIGURE 4–17. The lead for a single lead thread is the same as the pitch of the thread. The pitch is the distance from a point on one thread to the same point on the next thread.

The second type of tapping is called rigid tapping. Rigid tapping does not require special holders. Precise feed and RPM synchronization are needed, however, to insure undamaged threads. The feed of the tap needs to be calculated. The feed for tapping is calculated by dividing 1 (inch) by the number of threads per inch. The result is then multiplied by the revolutions per minute of the spindle. For example: 1/4-20 UNC tap running at 250 RPM. 1/20 * 250 = 0.05 * 250 RPM = 12.5 inches per minute feed rate. Most modern machining centers are equipped with tapping cycles, which feed the tap down to the programmed depth and then automatically reverse the spindle and feed up, unscrewing the tap from the hole.

Tools for Milling

When selecting the best milling cutter for a particular operation, four things must be taken into consideration; the kind of cut to be made, the material to be cut, the number of parts to be machined, and the type of machine available.

End Mills

One of the most frequently used tools on a machining center is the end mill (see Figure 4–18). End mills are made from two types of materials: solid carbide or high-speed steel. End mills are ground with a relief on the sides and ends just behind the cutting edges. They are available with two or more flutes. Two fluted or end-cutting end mills can be used for plunging. The teeth on the end are much like those of a drill.

Two considerations determine the number of flutes a milling cutter should have: does the end mill need to be capable of end cutting for a plunging operation, and what is the depth of cut going to be? Increasing the number of teeth on the end mill greatly reduces the chip clearance area and may result in ships clogging the end mill. End mills can be used for profile cutting, slotting, cavity cutting, or face milling. Face milling, however, is usually done with a face milling cutter. End mills are typically held in solid-type holders with set screws for positive holding (see Figure 4–19).

FIGURE 4–18. End mills are manufactured with two or more flutes. The two-flute double end mill is used for plunging and profiling. Ball end mills have a radius ground on the end of the tool and are used for milling radii in slots or contouring the bottom surfaces of mold cavities. The multi-fluted roughing end mill, or hog mill, has scallops or grooves around the body of the tool and can remove three times as much material as standard end mills.

FIGURE 4–19. End mill tool holders use set screws to positively locate on the flats that are located on end mills. This positive locking-style holder keeps the end mill from slipping during machining operations.

Face Milling Cutters

Face milling cutters are widely used because of their ability to take large facing cuts. Face mills range in size from 1-1/4 to 6 inches and up. They have a hole for mounting on an arbor and a keyway to receive a driving key (see Figure 4–20). Carbide face mills have carbide inserts that can be indexed or recycled when they become dull.

FIGURE 4–20. Small face milling cutter with indexable carbide inserts.

Climb and Conventional Milling

When milling there are two directions you can feed: into the rotation of the cutter or with the rotation of the cutter (see Figure 4–21). Feeding with the rotation of the cutter is known as *climb milling*. In climb milling, the cutter is attempting to *climb* onto the workpiece.

Climb milling is not typically done on manual machines unless they are equipped with backlash eliminators. If there is any slack or backlash between the screw and nut driving the table, the workpiece is pulled into the cutter, possibly causing tool breakage, a scrapped workpiece, and possible serious injury to the operator. CNC machines are equipped with ball screws which virtually eliminate backlash, thus allowing climb milling.

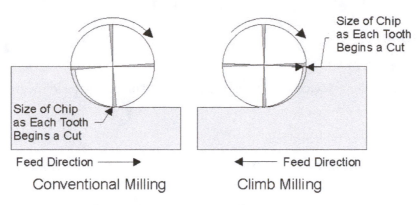

Conventional Milling   Climb Milling

FIGURE 4–21. Climb milling and conventional milling represent the two directions of feed associated with milling. When climb milling, the outer scale of the material is cut first. In conventional milling, the inside of the material is cut first.

Climb milling is desirable in most cases because it takes less horsepower. Other benefits of climb milling are better surface finishes, less tool deflection, extended tool life, and the chips are discarded away from the cutter.

Feeding against the rotation of the cutter is known as *conventional milling*. The conventional milling chip has no thickness at the beginning, but builds in size toward the exit of the cutter. Conventional milling is recommended on materials with a hard outer scale, such as cast steels or forgings. If the tool needs to extend out of the holder a greater-than-normal length, it may be best to conventional cut. This will cause the tool to flex and stay in the flexed position, avoiding chatter.

Chapter Questions

1. What type of work is typically held in a vise?

2. What type of work-holding device would you use to hold a workpiece at a 90-degree angle to the axis of travel?

3. How are large workpieces typically held?

4. What types of work-holding devices are typically used in high-volume-type jobs?

5. What is the most common hole-producing tool?

6. What keeps tapered-shank drills from slipping during drilling?

7. What is the purpose and advantage of center drills?

8. How does using spade drills reduce the cost of producing holes?

9. What is a reamer?

10. What tools are used for boring?

11. Describe the two types of tapping that are done on a machining center.

12. What type of end mill would be used for plunge cutting?

13. What is a carbide face milling cutter?

Chapter 5

FUNDAMENTALS OF PROGRAMMING

INTRODUCTION

CNC programming is the process of taking the information from a part print that you would use to do manual machining and converting it to a language that a CNC machine will understand. This chapter focuses on word address programming.

OBJECTIVES

Upon completion of this chapter, the reader will be able to:

- Explain the parts of a program.
- Arrange and explain blocks of information.
- Describe preparatory and miscellaneous functions.
- Describe G92 and G54 workpiece coordinate settings.
- Write simple programs using word address format.
- Explain two methods of programming an arc.
- Describe the use of tool height and tool diameter offsets.
- Write programs that use offsets.

Word Address Programming

Word address programming precisely controls a machine's movements and functions by using short sentence-like commands. Let's consider a simple operation that could be performed on a manual milling machine and convert it to a word address command: Turn the spindle on in a clockwise direction at a spindle speed (RPM) of 600.

The command to do this on a CNC machine would be: M03 S600. The M03 commands the spindle to start in a clockwise direction. The S600 tells the spindle how fast to turn. This is one block of information.

Letter Address Commands

A CNC machine is controlled by the use of *letter address commands*. Following are abbreviated descriptions of the most common letter address commands.

    N is used for the line number or sequence number for each block of program.

    G is used for specific modes of operation. G-codes are also called preparatory functions. G-Codes set up the mode in which the machining operation(s) are to be executed.

F is the feed rate.

S is the spindle speed setting.

T is the letter address for a tool.

M is a miscellaneous function. Miscellaneous functions include coolant on/off, spindle forward, spindle reverse, and many others.

H and D are letter address codes used for tool height and tool diameter offsets.

Programming Terminology

A program word is composed of two parts: an address and a number. Study Figure 5-1. In the first example, G is the address and 01 is the number. Together they are called a *word*. A G01 word commands the machine to make a linear move. The second example of a word (S800) is composed of the S address and 800. S800 would set a spindle speed of 800 RPM. The third example (M08) of a word is composed of an address of M and 08. An M08 word commands the control to turn flood coolant on.

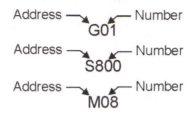

Figure 5-1. Examples of word addresses.

A block is one line of code for a CNC machine. Think of a block as being one operation. It can consist of one or more words (study Figure 5–2). Each block is ended with an end-of-block character (;). Note that the end-of-block character is usually generated automatically when the programmer hits the enter key at the end of a line. The end-of-block character does not normally show up on the machine display. When operating, a CNC control reads one block and executes it, it then reads the next block and executes it, and so on (see Figure 5–3).

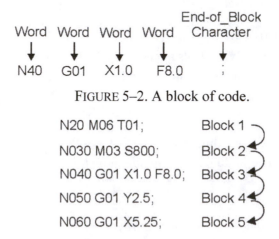

FIGURE 5–2. A block of code.

FIGURE 5–3. Order of execution.

The order of words in a block can vary. Figures 5–4 and 5-5 show the typical arrangement and order of words in blocks. Note that most blocks will contain far fewer words.

	N G X,Y,Z F S T M H,D
N	Sequence (line) number
G	Preparatory function. Specifies the mode of operation under which a command will be executed.
X,Y,Z	Dimension words. Used to specify a position or distance to move.
S	Spindle speed
T	Designates the tool to be used.
M	Miscellaneous function code. Designates things such as spindle on/off, spindle direction, coolant on/off, etc.
H,D	Used to specify tool offsets for height and/or diameter.

FIGURE 5–4. Typical arrangement of words in a block.

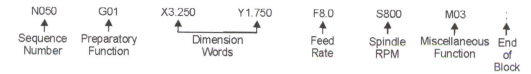

FIGURE 5–5. Typical order for one block of code.

Program Numbers

Programs are stored in the controller by their program number. Controls can store many programs. Program numbers start with the letter O. For example, O5 would be program 5. The program number normally appears before the first line of code in a program (see Figure 5–6). In this figure, the program number is O100. Note also the use of sequence numbers to name blocks of code. Remarks can be put in parentheses on any line. The controller ignores remarks.

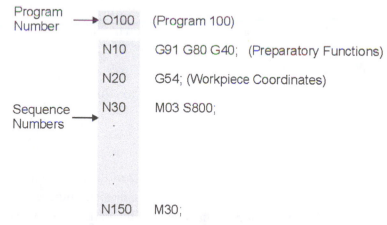

FIGURE 5–6. Use of program number and sequence numbers.

Part Programming

A part program is simply a series of blocks that execute motions and machine functions to make a part. Let's take a look at a simple program that cuts around the outside of a 3-inch by 4-inch block with a 1/2-inch diameter end mill (see Figure 5–7). We had to compensate for the radius of the cutter to make the part the correct size. The cutter radius is .250 so the program must keep the center of the tool .250 to the left of the part as it moves in a clockwise direction around it.

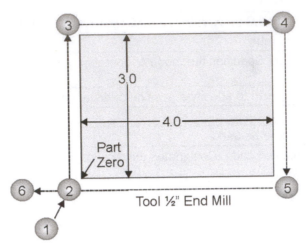

FIGURE 5–7. Profile mill of a 3-inch by 4-inch block.

```
N0010 M03 S800
N0020 G00 X-1.00 Y-1.00 (point 1)
N0030 G01 X-.25 Y-.25 F10.0 (point 2)
N0040 G01 Y3.25 (point 3)
N0050 G01 X4.25 (point 4)
N0060 G01 Y-.25 (point 5)
N0070 G01 X-.50 (point 6)
```

N0010 M03 S800

Line number 10 turns on the spindle in a clockwise direction at 800 RPM.

N0020 G00 X-1.00 Y-1.00

Line number 20 is a rapid feed move (G00) to position the tool just off the lower left-hand corner of the part (X-1, Y-1). This is position 1 in Figure 5–7.

N0030 G01 X-.25 Y-.25 F10.0

Line number 30 feeds (G01) the tool to a position that is 1/2 of the tool diameter to the left of the side of the part (point 2, X-.250). The tool will feed at 10 inches per minute (F10.0 IPM). The right edge of the cutter is now aligned with the left-hand side of the part. It is now ready to cut the left side of the part.

N0040 G01 Y3.25

Line number 40 cuts the left side of the part and positions the spindle center past the top of the part (Y3.25) by 1/2 the tool diameter (point 3). This positions the edge of the tool for the cut across the top of the part. The feed rate is still 10 inches per minute.

N0050 G01 X4.25

Line 50 cuts the top of the part. The feed rate is still 10 IPM, because we have not changed it since line number 20. The spindle center is now positioned .25 to the right of the part (point 4). This gets us ready to cut the right-hand side of the part.

N0060 G01 Y-.25

Line number 60 moves the tool to point 5. The right-hand side of the part is now complete. The tool center is also positioned 1/2 tool diameter below the bottom of the part, ready to cut.

N0070 G01 X-.50

Line number 70 cuts the bottom of the part to size and moves the tool completely off the part (point 6, X-.50).

Next let's take a closer look at the individual parts of a word address part program.

Part Datum Location

To program a part, you need to determine where the workpiece zero (or part datum) should be located. The part datum is a feature of the part from which the majority of the dimensions of the part are located.

Because all of the dimensions of the part in Figure 5–7 come from the lower left-hand corner, this was the logical choice for the workpiece zero point.

It is good practice to choose a part feature that is easy to access with an edge finder and one that will involve the fewest number of calculations. This approach will help avoid errors.

Sequence Numbers (Nxxxx)

Sequence numbers are a way to identify blocks of information within a program. Line numbers are handy for the operator. The machine controller can be commanded to find blocks of information by their line numbers.

In addition, line or sequence numbers are needed in the use of some canned cycles, which will be covered later in this book.

G-Codes (Preparatory Functions)

G-codes are used to set modes such as linear interpolation (G01) or rapid traverse (G00). Linear interpolation means that the cutter moves on a controlled linear path (line). A "G" followed by a two-digit number determines the machining mode in that block or line.

G-codes or preparatory functions fall into two categories: modal or nonmodal. Nonmodal or "one-shot" G-codes are those command codes that are only active in the block in which they are specified.

Modal G-codes are those command codes that will remain active until another code of the same type overrides it.

For example, if you had five lines that were all linear feed moves, you would only have to put a G01 in the first line. The other four lines would be controlled by the previous G01 code. The feed rate was modal in the first example. The feed rate does not change unless a different feed rate is commanded.

The G-codes shown in Figure 5–8 are commonly used machining center G-codes, but some of them may be slightly different from those used on your machine tool. Consult the manual for your machine to be sure. Note: Appendix A has a more extensive list of common G codes and examples of their use.

G00	Rapid traverse (rapid move)	Modal
G01	Linear positioning at a feed rate	Modal
G02	Circular interpolation clockwise	Modal
G03	Circular interpolation counter-clockwise	Modal
G17	XY Plane Selection	Modal
G20	Inch Programming	Modal
G28	Zero or home return	Non-Modal
G40	Tool diameter compensation cancel	Modal
G41	Tool diameter compensation-left	Modal
G42	Tool diameter compensation-right	Modal
G43	Tool height offset	Modal
G49	Tool height offset cancel	Modal
G54	Workpiece coordinate preset	
G80	Canned cycle cancel	Modal
G81	Canned drill cycle	Modal
G83	Canned peck drill cycle	Modal
G84	Canned tapping cycle	Modal
G85	Canned boring cycle	Modal
G90	Absolute coordinate positioning	Modal
G91	Incremental positioning	Modal
G92	Workpiece coordinate preset	
G98	Canned cycle initial point return	Modal
G99	Canned cycle R point return	Modal

FIGURE 5–8. Commonly used machining center G-codes.

Spindle Control Functions

Spindle speeds are controlled with an "S" followed by up to four digits. When programming a machining center, the spindle speed is programmed in revolutions per minute (RPM). A spindle speed of 600 RPM would be programmed S600. Spindle speeds may also be programmed in surface feet per minute (SFPM) through the use of a G96 preparatory code. SFPM is the cutting speed of the material you are machining. Most turning center controls will typically program in SFPM. This allows the spindle speed to automatically change as the diameter of the workpiece changes, maintaining a constant surface speed. For example, a cutting speed for mild steel and a carbide tool might be 400 SFPM. A spindle speed of 400 SFPM would be programmed G96 S400. Then as the diameter of the turned part gets smaller, the RPM would increase to keep the cutting speed correct. The spindle is turned on using a miscellaneous (M) code of either M03 or M04. An M03 will turn the spindle on in a clockwise direction, while an M04 will turn the spindle on in a counterclockwise direction. A M05 turns the spindle off.

Miscellaneous Functions (M-Codes)

Miscellaneous functions or M-codes perform miscellaneous functions such as tool changes, coolant control, and spindle operations. An M-code is a two- or three-digit numerical value preceded by a letter address, "M." M-codes, similar to G-codes, can be modal or nonmodal. Figure 5–9 lists commonly used machining center miscellaneous functions (M-codes). Note: Appendix A has a more extensive list of common M codes and examples of their use.

M00	Program stop	Non-Modal
M01	Optional stop	Non-Modal
M02	End of program	Non-Modal
M03	Spindle start clockwise	Modal
M04	Spindle start counterclockwise	Modal
M05	Spindle stop	Modal
M06	Tool change	Non-Modal
M07	Mist coolant on	Modal
M08	Flood coolant on	Modal
M09	Coolant off	Modal
M30	End of program & reset to the top of program	Non-Modal
M40	Spindle low range	Modal
M41	Spindle high range	Modal
M98	Subprogram call	Modal
M99	End subprogram & return to main program	Modal

FIGURE 5–9. Miscellaneous functions.

Tool Calls

Tool calls are straightforward, although machining center tool calls are slightly different than turning center tool calls. On a machining center, a tool change is commanded with a miscellaneous code of M06. Next you tell the control which tool to change to. A typical tool change block for a machining center would be M06 T02. This line of code would tell the CNC controller to change to tool 2.

On a turning center, the tool call also starts with a "T" and the tool number (T02), but then you add the tool offset number. T0202 would be the tool call for tool number 02 with an offset of number 02. It is written 02 because you typically have more than 10 tools and 10 offsets available to you, for example, T1212 (tool 12, offset 12). The offset lets the operator correct any errors in the size of the part. An M06 is not used when programming a turning center to do a tool change. An M06 on the turning center is used to unclamp the chuck.

Axes Words (X, Y, Z)

On a typical milling machine we have three axes: X, Y, and Z. Each axis is specified by a letter (X, Y, or Z), which may be preceded by a direction sign (+ or -).

Chapter 1 covered the Cartesian coordinate system, which specifies how the axes of the machines are oriented as well as the direction of travel. A simple command block to rapid position the tool of the milling machine to 1 inch above the workpiece zero would be: N0010 G00 Z1.00. The G00 means a rapid move, and the Z1.0 means 1 inch above the workpiece zero.

There are additional axes on some machines. Figure 5–10 shows additional axes of motion. If a machine has additional auxiliary axes that can move in the X, Y, and Z directions, they are called U, V, and W. Rotational extra axes around the X, Y, and Z are labeled A, B, and C.

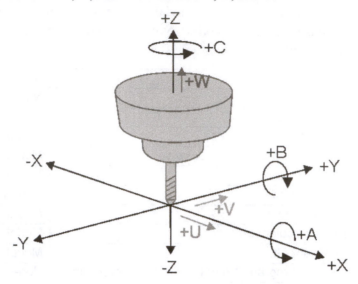

FIGURE 5–10. Machine axes of motion.

Motion Blocks
Motion can be controlled in three ways: rapid positioning, linear feed, or circular feed.

Rapid Traverse Positioning (G00)
A rapid positioning block consists of a preparatory G-code (G00) and the coordinate to which you want to move. A rapid move to a location of X10, Y5, and Z1 would be programmed: G00 X10.0 Y5.0 Z1.0. This block would command the machine to move at a rapid traverse rate to this position, moving all of the axes simultaneously. The rapid traverse rate for each machine is different, but it normally ranges from 100 to 600 or more inches per minute. The rapid traverse rate can usually be overridden using the manual rapid traverse over-ride switch located on the control. This means that an operator can choose to reduce the rapid rate from 0 to 100 percent. This should be done when testing a new program for the first time.

Linear Feed Mode (G01)
A G01 (or G1) linear interpolation code moves the tool to a commanded position in a straight line at a specific feed rate. The feed rate is the speed at which the machine axes move. Linear feed blocks are normally cutting blocks. The rate at which the metal is removed is controlled through a feed rate code (F). Machining centers can use feed rates in inches per minute (IPM), or inches per revolution of the spindle (IPR). The feed rate type selection of IPM or IPR is controlled by G codes.

Turning centers are typically programmed in inches per revolution of the spindle, or IPR. To make a straight-line cutting motion on a machining center, the block of information would look like this: G01 X10.00 F8.00. The tool would move to an X-axis position of 10.00 inches at a feed rate of 8 inches per minute. Remember, straight-line moves can also be angular. CNC machine controls are capable of making simultaneous axes moves (see Figure 5–11).

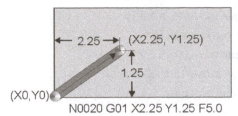

N0020 G01 X2.25 Y1.25 F5.0

FIGURE 5–11. G01 linear interpolation example.

The G00, G01, and F codes are all modal. Modal commands are active unless changed by another preparatory code. If you were programming a series of straight-line moves, you would only have to put the G01 and the feed rate in the first line. The lines that follow would be controlled by the previous G01 and feed rate. Note that it is OK to put the G01 or G1 in every line and it may make the program easier to understand.

The example shown in Figure 5–12 incorporates the programming procedures that have been covered to this point. We need to mill around the outside of the part, .250 inches deep, with a .50-inch end mill at a feed rate of 5 inches per minute. The first step is to set up the control by doing our preliminary procedures. The second step is the tool call. The third step sets the WPC or workpiece zero point. In our fourth step, we need to start the spindle and set the RPM. Next we rapid position close to the part and start our linear cutting moves. After we have cut the profile of the part, we need to return to the home position and end the program.

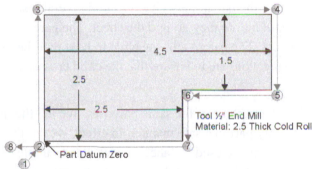

FIGURE 5–12. Simple contour programming example.

O001 (Program name O0001)

N100 G20 G17 G40 G49 G80 G90 (Preparatory Information - G20 - input data in inches, G17 - X, Y plane, G40 - cancel cutter diameter compensation, G49 - cancel tool length compensation, G80 - cancel fixed cycles, G90 - absolute mode)
 N101 G54 (Work piece fixture location 1)
 N104 T1 M6 (T1- Tool 1.5" end mill, M6 - tool change)
 N105 S800 M3 (S800 - Spindle speed 800, M3 - spindle on clockwise)
 N106 G00 X-1.0 Y-1.0 (G00 - Rapid move to X-1.0 Y-1.0)
 N108 G00 Z.1 M8 (G00 - Rapid to Z. 1, M8 - flood coolant on)
 N110 G01 Z-.25 F5.0 (G01 - linear move to Z-.25, F5.0 - Feed rate 5 inches per minute)
 N111 G01 X-.25 Y- .25 (G01 - Linear move to X-.25 Y-.25 at same feed rate as the previous line)
 N114 G01 Y2.75 (G01 - Linear move to Y2.75)

N116 G01 X4.75 (G01 - Linear move to X4.75)
N118 G01 Y.75 (G01 - linear move to Y.75)
N120 G01 X2.75 (G01 - Linear move to X2.75)
N122 G01 Y-.25 (G01 - Linear move to Y-.25)
N124 G01 X-1.00 (G01 - Linear move to X-1.00)
N132 G00 Z.1 (G00 - Linear move to Z.1)
N134 M05 (M05 - Spindle stop)
N136 G28 M09 (G28 - Return to reference point, M09 - flood coolant off)
N138 M30 (Program end, memory reset)
%

Programming

Programs can be thought of as having three sections. In fact, a CNC program is similar to a business letter. There is a heading, the body of the letter, and then an ending.

- The heading, or the introduction, consists of preparatory codes.
- The body consists of the machining operations.
- The ending consists of codes needed to end the program.

Preparatory Codes

Preparatory codes are used to set conditions or cancel conditions. When a parent with a child approaches a corner and wants to cross a street, the parent may give the child several instructions. The parent tells the child to stop running and stand still before they go into the street. The parent tells the child to look both ways before stepping into the street. The parent tells the child to hold their hand and walk across the street. The instructions given to the child are to ensure that the child crosses the street safely and successfully.

Preparatory functions for a CNC have some of the same purposes. We tell the machine how we want it to operate in order to produce a quality part in a safe manner. In effect, we tell the machine how we want it to operate. Some preparatory G-codes are used to cancel modes. For example, if we just finished running a program that used offsets, we would like to cancel any offsets that may be in memory.

A G40 would cancel diameter offsets and a G49 would cancel height offsets. We would like to cancel any canned cycles that may have been active. So those are two examples of canceling modes that may be active for safety and proper operation. Other codes are used to set the mode we want to operate in. For example, if our program is written in inch mode (not metric), we would use a G20. If our program is developed in the XY plane, we would tell the control to use the XY plane by using a G17.

Study the lines of code below. There are three lines. Each line of code is considered one block of information to the control. The first line is O0001. This is the name of the program. Line N10 has a G20 code. This tells the control that our program is written in inch mode (not metric). Study line N020. The G17 tells the control that the program was written in the XY plane. The G40 tells the control to cancel diameter offsets. The G49 cancels height offset. The G80 cancels canned cycles. The G90 sets the control to absolute mode.

O0001
N10 G20
N20 G17 G40 G49 G80 G90

The order in which they are done and the line numbers are unimportant. In fact, many shops have somewhat standard header files they use to cancel all offsets and canned cycles and set the modes they normally use for programming.

Codes to Consider for Preparatory Functions

Codes to cancel modes:

G40 Cancel diameter offset

G49 Cancel height offset

G80 Cancel canned cycles

Codes to set modes of operation:

G17 XY plane designation

G18 ZX plane selection

G19 YZ plane selection

G20 Inch mode

G21 Metric mode

G54 Workpiece coordinates 1 (Many machines also allow G55-G59 to be used for alternative workpiece coordinates.)

G90 Absolute programming

G91 Incremental programming

G94 Feet-per-minute

G95 Feed-per-revolution

Comments

Other information may also be included in the program. Many companies include setup information for the operator. Study the example that follows. Note the parentheses around the comments in each block. The control ignores text that is between parentheses. It is only for operator information.

O0002
N010 (X0 Y0 is the lower left-hand corner)
N020 (Tool 1: 1" spot drill)
N030 (Tool 2: .25 drill)
N040 (Tool 3: .500 end mill)
N050 (Set the part up in a vise with 2 inch parallels)

Machining Operations

The body of the program contains the machining operations. The code in the body of the program is used to load the proper tools, control speeds, feeds, and motion of tools.

End the Program

This is the simplest part of the program. In this part of the program, you might want to make sure the tools have been put away, cancel canned cycles, and tell the control the program is complete.

Workpiece Coordinate (WPC) Setting

The machine must know where the part is on the table. Remember that the machine has a zero position. Figure 5–13 shows that for this machine, the machine zero position is located at the left front and top of the machine table. The machine always knows where this is. The machine has no idea where the workpiece is until we tell it where it is. The part position can be called workpiece coordinates. This is called workpiece coordinate setting. The WPC tells the machine the position of the part datum (see Figure 5–14).

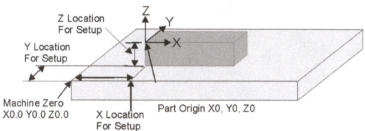

FIGURE 5–13. Workpiece coordinates.

The workpiece part zero may be located at the corner or any other part of the workpiece, but we have to tell the control where this point is on the machine table. The technique for locating the workpiece zero varies for each machine tool. Some controls use a button to set the zero point. The setup person or operator uses the jog buttons to position the spindle center over the part datum and then presses the zero set or zero shift button to set the coordinate system to zero. On other types of controls, the WPC is set with a G-code. There are two common WPC G-codes: G54 and G92. Machines use either a G54-G59 for stored values or a G92 followed by X, Y, and Z dimensions.

The G54-G59 workpiece coordinate is the absolute coordinate position of the part zero (see Figure 5–14). These are not available on all machines. Six are available, and all serve the same function. This allows the programmer to have six different workpiece coordinates established on a machine. This would be very beneficial for repetitive jobs that could be located at the same position on the machine table. For example, a job that is run once each week might use the G59. The G59 would be used to establish the location for that particular job.

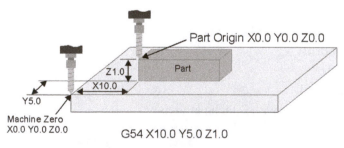

FIGURE 5–14. Workpiece coordinates.

To locate this position, the operator would position the center of the spindle directly over the part zero using an edge finder or probe and then take note of the machine position (see Figures 5–15 and 5–16). The coordinates of this position would be placed in the G54 line. A typical workpiece coordinate setting of this type would be written: N0010 G54. Note that the values for X, Y, and Z are stored in the CNC and don't appear in the program.

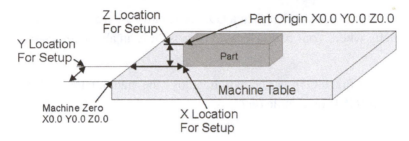

FIGURE 5–15. Machine Zero and part origin.

FIGURE 5–16. Locating the corner of the part using an edge finder. An edge finder or probe can be used to precisely position the center of the spindle over the part datum (zero).

The G92 workpiece coordinate is the incremental distance from the workpiece datum (X, Y, and Z zero) to the center of the spindle (see Figure 5–17). In effect, it tells the machine where the spindle is in relation to the workpiece. The spindle must then be in that position when you start to run the program.

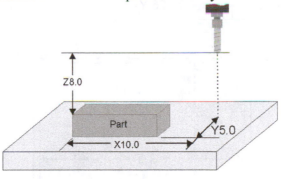

G92 -X10.0 Y-5.0 Z-8.0

FIGURE 5–17. G92 workpiece coordinates.

When the G92 is called, the center of the spindle has to be in the preprogrammed position. If it is not, the control will start machining at the wrong position. If, for example, the center of the spindle at the home position is 10 inches to the right on the X axis, 5 inches back on the Y axis, and 8 inches above the part on the Z axis, the G92 would be written: G92 X10.00 Y5.00 Z8.00. If the center of the spindle were located any other distance away from the part datum, when the G92 was called, the tool would try to cut the part in the wrong location. This is why using a G92 can be very dangerous!

The G54 type of WPC setting is a lot safer than a G92. No matter where you are when the G54 is called, the control knows exactly where your part is located because it is an absolute position, not an incremental distance.

It is important to remember that a G54 or G92 will not move the machine tool to this point; it merely tells the control where the part (G54) or spindle (G92) is.

Incremental Programming

Now that we have established the basics of absolute programming, we can look at another type of positioning, incremental positioning. Absolute programming is when all of the coordinates of the part program are related to an absolute zero point. Incremental programming defines the coordinates of the part in relationship to the present position.

Incremental programming is also known as point-to-point positioning. The point where you are presently is the zero for the next coordinate position. Incremental positions are the direction (+ or −) and the distance to the next point. Study Figure 5-18 which contrasts absolute and incremental moves. For example to move from point 3 to point 4 in incremental mode X axis would not need to move and Y would need to move -3.5". Incremental moves are simply the distance and direction needed to go from the current machine position to the next position.

Tool Movement	Absolute Command	Incremental Command
Point 1 to Point 2	X-.25 Y3.25	X0.0 Y3.5
Point 2 to Point 3	X4.25 Y3.25	X4.5 Y0.0
Point 3 to Point 4	X4.25 Y-.25	X0.0 Y-3.5
Point 4 to Point 1	X-.25 Y-.25	X-4.5 Y0.0

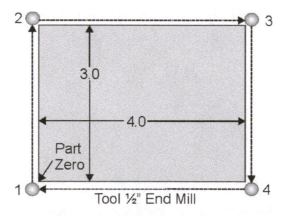

FIGURE 5–18. Incremental positioning.

This type of programming can be used to program the whole part or just certain sections of the program. Incremental positioning is programmed with a G91 preparatory code and can be very useful when programming a series of holes that are incrementally located on the part print. Figure 5–19 would be a typical application for incremental programming.

Most of the program to drill the holes could just be incremental .75" moves in the X or Y direction. If you wished to switch back to absolute programming at any point in the program, you would use a G90 code.

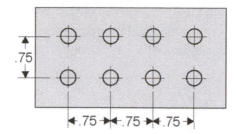

FIGURE 5–19. Incremental programming example.

Circular Interpolation

Up to this point, we have discussed only straight-line moves. If a CNC machine was only capable of straight-line moves, it would be very limited. One of the most important features of a CNC machine is the ability to do circular cutting motions. CNC machines are capable of cutting any arc of any specified radius value. Arc or radius cutting is known as *circular interpolation.* Circular interpolation is carried out with a G02 or G03 code.

Programming Circular Moves

There are two basic methods used to program circular moves: incremental arc center or radius.

Programming Circular Moves Using the Incremental Arc Center Method

When we start cutting an arc, the tool is already positioned at the start point of the arc. First, we need to tell the direction of the arc. Is it a clockwise (G02) or counterclockwise (G03) arc? The second piece of information the control needs is the end point of the arc. The last piece of information is the location of the arc center (see Figure 5–20).

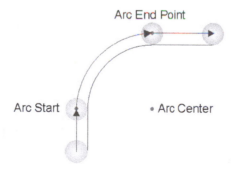

FIGURE 5–20. The critical pieces of information needed to cut an arc are the arc start point, arc direction, arc end point, and arc centerpoint location.

Arc Start Point

The arc start point is the coordinate location where the arc starts. The tool is moved to the arc start point in the line prior to the arc generation line. Simply stated, the start point of the arc is the point where you are when you want to start machining the arc (see Figure 5–20).

Arc Direction (G02, G03)

Circular interpolation can be carried out in two directions, clockwise or counterclockwise. There are two G-codes that specify arc direction (see Figure 5–21). The G02 code is used for circular interpolation in a clockwise direction. The G03 code is used for circular interpolation in a counterclockwise direction. Both G02 and G03 codes are modal and are controlled by a feed rate (F) code, just like a G01.

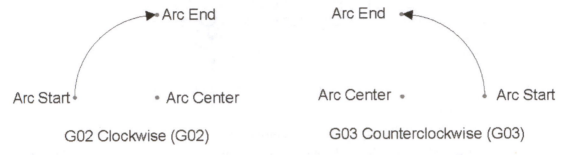

FIGURE 5–21. G02 and G03 arc direction.

Arc End Point

The arc end point is the coordinate position for the end point of the arc. The arc start point and arc end point set up the tool path, which is generated according to the arc center position (see Figure 5–22 and 5-23).

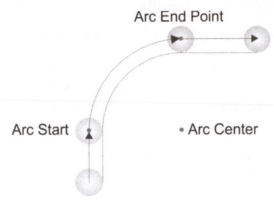

FIGURE 5–22. Arc end point location.

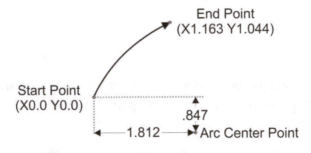

G02 X1.163 Y1.044 I1.812 J-.847

FIGURE 5–23. The center point of the arc is the distance from the start point position to the arc center position.

Arc Centerpoints

To generate a circular path, the controller has to know where the center of the arc is. When using the IJ method, a particular problem arises. How do we describe the position of the arc center? If we use X, Y, Z coordinate position words to describe the end point of the arc, how will the controller discriminate between the end point coordinates and the arc center coordinates? The answer is that we use different letters to describe the same axes. Secondary axes addresses are used to designate arc centerpoints. The secondary axes addresses for the axes are:

I = X axis coordinate of an arc center point

J = Y axis coordinate of an arc center point

K = Z axis coordinate of an arc centerpoint

Because we will be cutting an arc in only two axes, only two of three secondary addresses will be used to generate an arc. When cutting arcs in the X/Y axes, the I/J letter addresses will be used. If we were cutting an arc on a turning center, the X/Z axes would be the primary axes, and the I/K letter addresses would be used to describe the arc centerpoint.

The type of controller you are using dictates how these secondary axes are located. With most controllers, the arc center point position is described as the incremental distance from the arc start point to the arc center (see Figure 5–23).

While very uncommon, there are a few controls where the arc centerpoint position is described as the absolute location of the arc centerpoint from the workpiece zero point (see Figure 5–24). Note: this is very uncommon.

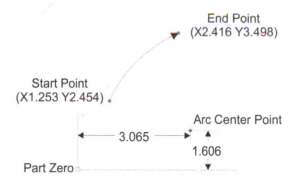

G02 X2.416 Y3.498 I3.065 J1.606

FIGURE 5–24. When using absolute center point positioning, the center point is the coordinate position from the part zero. This method is very uncommon.

If the arc center point is located down or to the left of the start point, a negative sign (–) must precede the coordinate dimension (see Figure 5–25).

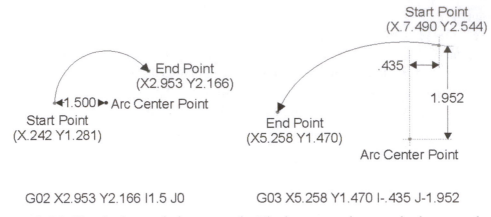

G02 X2.953 Y2.166 I1.5 J0   G03 X5.258 Y1.470 I-.435 J-1.952

FIGURE 5–25. Circular interpolation example. The incremental arc method was used.

Remember that I and J values are normally incremental. I and J values must be the incremental distance and direction. The direction is shown by the plus or minus sign. Also remember that the distance and direction is taken from the start of the arc to the center of the arc.

Circular Interpolation Using the Radius Method

Circular motion can also be programmed using the radius of the arc. The general format is G02 (or G03) X Y R. Figure 5–26 shows an example of a circular cut using the radius method. The code for this cut would be G02 X4.000 Y3.000 R1.0. The G02 means that it will be a clockwise circular cut. The X value is 4.000 inches (absolute) in the X axis. The Y value is 3.000 inches (absolute) in the Y axis. The R value is the length of the radius value. In this example, it is +1.0 inches.

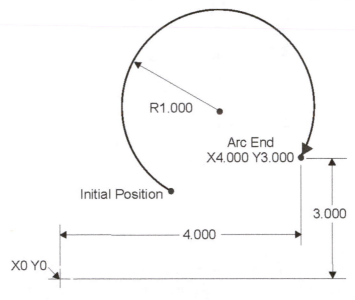

FIGURE 5–26. Circular cut using the radius programming method.

Study Figure 5–27. Note that there are two possible arcs with the same start point, same end point, and radius of 1.0". One is greater than 180 degrees and the second one is less than 180 degrees. We tell the controller which one to use by giving the radius value a plus or minus sign.

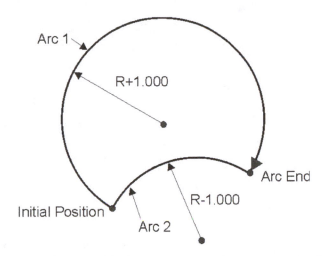

FIGURE 5–27. Two possible arcs with the same start point, same end point, and same radius.

Figure 5–28 shows an arc that is more than 180 degrees. The code to machine this arc would be G02 X4.0625 Y3.0625 R-2.5. G02 means that it is a clockwise arc. The X value is the absolute X position of the end point. The Y value is the absolute position of the end point in the Y axis. The R value is the value of the radius of the arc. Note that it must be negative in this example because the arc is more than 180 degrees.

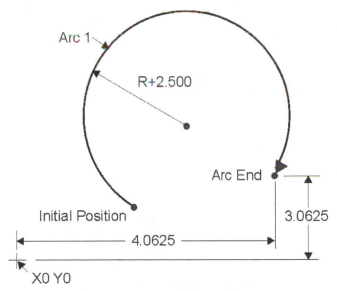

FIGURE 5–28. Arc of more than 180 degrees.

Figure 5–29 shows an arc that is less than 180 degrees. The code to machine this arc would be G02 X4.0625 Y3.0625 R2.5. G02 means that it is a clockwise arc. The X value is the absolute position of the end point in the X axis. The Y value is the absolute Y position of the end point. The R value is the value of the radius of the arc. Note that it must be positive in this example because the arc is less than 180 degrees.

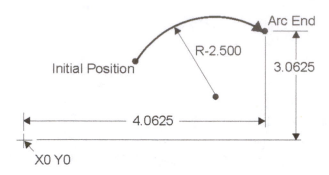

FIGURE 5–29. Arc of less than 180 degrees.

The radius method cannot be used to program a full circle. The IJ method should be used for full circles.

Comprehensive Programming Exercise

Figure 5–30 shows a base plate that involves linear and circular cutting. To simplify the programming of the part, the holding device has been eliminated and programming is done at the center of the spindle (no tool offsets).

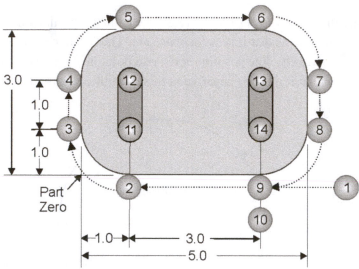

FIGURE 5–30. Base plate.

O0001 (Program name O0001)
N10 T2 M6 (Tool change to tool 2)
N20 G54 (Workpiece coordinates are in G54 register)
N30 M3 S800 (Turn spindle on clockwise direction 800 RPM)
N40 G00 X6.00 Y – .25 (Rapid position spindle to X6.0 Y-.25)
N50 G00 Z.100 (Rapid position to Z.100)
N60 G01 Z – .500 F6.0 (Linear feed to Z-.500 at 6.0 IPM)
N70 G01 X1.00 (Linear feed to X1.00 at 6.0 IPM)
N80 G02 X – .25 Y1.00 I0.0 J1.25 (Circular mill an arc in the clockwise direction at 6.0 IPM)
N90 G01 Y2.00 (Linear feed to Y2.00 at a feed rate of 6.0 IPM)
N100 G02 X1.00 Y3.25 I1.25 J0.0 (Circular mill an arc in the clockwise direction at 6.0 IPM)
N110 G01 X4.00 (Linear feed to X4.00 at 6.0 IPM)

N120 G02 X5.25 Y2.00 I0.0 J – 1.25 (Circular mill an arc in the clockwise direction at 6.0 IPM)

N130 G01 Y1.00 (Linear feed to Y1.00 at 6.0 IPM)

N140 G02 X4.00 Y – .25 I – 1.25 J0.0 (Circular mill an arc in the clockwise direction at 6.0 IPM)

N150 G01 Y – 1.00 (Linear feed to -Y1.00 at 6.0 IPM)

N170 M05 (Turn spindle off)

N180 M6 T1 (Tool change to tool 1)

N190 M03 S750 (Spindle on counterclockwise at 750 RPM)

N200 G00 X1.00 Y1.00 (Rapid to X1.00 Y1.00)

N210 G00 Z.100 (Rapid to Z.100)

N220 G01 Z – .525 F2.0 (Feed to Z-.525 at 6.0 IPM)

N230 G01 Y2.00 F5.0 (Feed to Y2.00 at 5.0 IPM)

N240 G01 Z.100 (Feed to Z.100 at 5.0 IPM)

N250 G00 X4.00 (Rapid to X4.00)

N260 G01 Z – .525 F2.0 (Feed to Z-.525 at 2.0 IPM)

N270 G01 Y1.00 F5.0 (Feed to Y1.00 at 5.0 IPM)

N280 G01 Z.100 (Feed to Z.100 at 5.0 IPM)

N300 T0 M6 (Put the tool away)

N310 M30 (Program end, memory reset)

%

Tool Length Offsets

Up to this point, we have not looked closely at tool length offsets. Length offsets make it possible for a CNC machine to adjust to different tool lengths. Every tool is going to be a different length, but CNC machines can deal with this quite easily. CNC controllers have a special area within the control to store tool length offsets.

The tool length offset is the distance from the tool tip at home position to the workpiece Z zero position (see Figure 5–31). This distance is stored in a table that the programmer can access using a G-code or tool code. On a machining center a G43 code is typically used. The letter address G43 code is accompanied by an "H" auxiliary letter and a two-digit number. The G43 tells the control to compensate the Z axis, while the H and the number tell the control which offset to call out of the tool offset table. The tool length offset typically needs to be accompanied by a Z axis move to activate it.

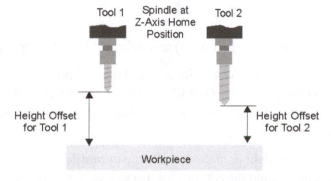

FIGURE 5–31. Tool length offsets.

A typical tool length offset block would look like this:

    N0010 G43 H10;

The G43 calls for a tool length offset, and the H10 is the number of the offset, which is found in register 10 of the tool length offset file. It is a good idea to correspond the tool length offset register number to the tool number. For example, if you are using tool number 10 (T10), try to correspond the height offset by using height offset number 10 (H10).

On some other brands of CNC machines, the height offset is called up with the tool number. If the program calls for tool number 10, the control automatically accesses the tool file and offsets the tool according to the tool length registered in the tool file under the tool number 10.

Because machine controls vary, it is a good idea to find out how your control deals with variations in tool lengths.

Tool Diameter Offsets

Tools also differ in diameter, and to compensate for this we use tool diameter offsets. Tool diameter offsets are also used to control the size of milled features. Tool diameter offsets allow you to program the part, not the toolpath.

In Figure 5–32 we had to compensate for tool diameter by offsetting the tool path by the radius of the tool to the left or right. The control can offset the path of the tool automatically so we can program the part just as it appears on the part print. This saves us from having to mathematically calculate the cutter path. The diameter offset also allows the programmer to use the same program for any size cutter by just changing the offset. Without diameter offsets, the programmer would have to know the precise size of the tool to be used and program using the center of the spindle.

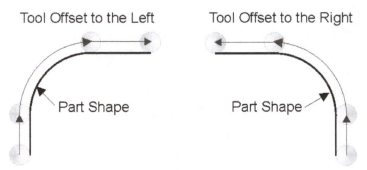

FIGURE 5–32. Left (G41) and right (G42) tool compensation.

With cutter compensation, the cutter size can be ignored and the part profile can be programmed. The radius of the cutting tool is entered into the offset file, and when the offset is called, the tool path will automatically be offset by the tool radius. If the part is too big or too small, the offset can be changed so that the next part will be accurate.

Cutter compensation can be to the right or to the left of the part profile. To determine which offset you need, imagine yourself walking behind the cutting tool. Do you want the tool to be on the left of the programmed path or to the right (see Figure 5–32)?

When compensation to the left is desired, a G41 is used. When compensation to the right is desired, a G42 is used. When using the cutter compensation codes, you need to tell the controller which offset to use from the offset table. The offset identification is a number that is placed after the direction code. A typical cutter compensation line would look like this: G41 D12.

To initialize cutter compensation, the programmer has to make a move (ramp on). This additional move must occur before cutting begins. This move allows the control to evaluate its present position and make the necessary adjustment from centerline positioning to cutter periphery positioning. This move must be larger than the amount of the tool offset. The machine corrects for the offset of the tool during the ramp move. In Figure 5-33 the machine compensates for the offset in the move between point 1 and point 2.

To cancel the cutter compensation and return to cutter centerline programming, the programmer must make a linear move (ramp off) to invoke a cutter compensation cancellation (G40). This is an additional move after the cut is complete (point 5 to point 6). Do not cancel the cutter compensation until the tool is away from the part by at least as much as the compensation amount. Figure 5–33 shows a typical programming example that uses tool length and tool diameter compensation.

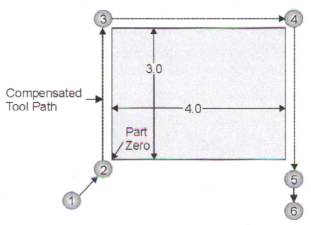

FIGURE 5–33. Programming example using tool length and cutter diameter compensation.

N10 G17 G20 G40 G80 G90 (G17 - X, Y plane, G20 - input data in inches, G40 - Cancel Cutter diameter compensation, G80 - cancel fixed cycles, G90 - Absolute mode)

N20 G54 (Work piece coordinates 1)

N30 T2 M06 (Tool change to tool 2- .375 end mill)

N40 S800 M03 (Turn spindle on clockwise at 800 RPM)

N50 G00 X-1.0 Y-1.0 (Rapid to X-1.0 Y-1.0 Position 1)

N60 G43 H2 Z.1 (Tool length compensation positive by the amount in H2, rapid to Z.1)

N70 G01 Z- .5 F5.0 (Linear move to Z-.5 at a feed rate of 5.0 inches per minute)

N80 G41 D2 X0.0 Y0.0 (Cutter diameter compensation left by the amount found in D2, Ramp compensation on during linear move to X0.0 Y0.0 Position 2)

N90 G01 Y3.0 (Linear move to Y3.0 Position 3)

N100 G01 X4.0 (Linear move to X4.0 Position 4)

N110 G01 Y-.50 (Linear move to Y-.25 Position 5)

N260 G40 Y-1.0 (Ramp off and cancel cutter compensation feed to position 6)

N270 G00 Z.1 (Rapid to Z.1)

N280 M5 (Spindle stop)

N290 T0 M06 (Return to tool change position)

N300 M30 (Program end, memory reset)

%

It is very important that the programmer's tooling intentions are communicated to the operator. This is usually done using the part plan and setup sheets. These will be examined in the following chapters.

Chapter Questions

1. What is the most common CNC language in use today?
2. What primary role do preparatory functions serve?
3. Name three functions that miscellaneous codes control.
4. Name two considerations that must be taken into account when selecting a part datum location.
5. What does modal mean?
6. Complete the following table.

Code	Function
G00	
G01	
M03	
G54	
G92	
G02	
G03	
G70	
M08	

7. Write a line of code to move the spindle 10 inches to the right in the X axis at a feed rate of 10 inches per minute.
8. Write a line of code to move the Y axis 5 inches in a negative direction at a feed rate of 10 inches per minute.
9. Write a line of code to turn the spindle on in a clockwise direction at 800 RPM.
10. Write a line of code to make sure the control is in absolute and inch mode.
11. Write a line of code to change to tool 4.
12. Write a line of code to move the Z axis up 5 inches in rapid mode.
13. Describe the difference between G92 and G54 workpiece coordinate settings.
14. Write code to make the following move at a feed rate of 10 inches per minute. Write it in absolute mode. Move to X8.000 Y5.250
15. Write code to make the following move at a feed rate of 10 inches per minute. Write it in incremental mode. Move 5.000 inches to the right.
16. Write code to make the following move at a feed rate of 10 inches per minute. Write it in absolute mode. Move to X-7.500 Y-3.250.
17. Write a line of code using the incremental arc center method to machine the following arc at a feed rate of 5 inches per minute.

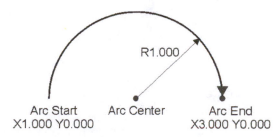

18. Write a line of code using the incremental arc center method to machine the following arc at a feed rate of 8 inches per minute.

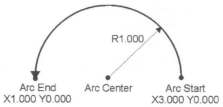

Arc End
X1.000 Y0.000

Arc Center

R1.000

Arc Start
X3.000 Y0.000

19. Write a line of code using the incremental arc center method to machine the following arc at a feed rate of 5 inches per minute.

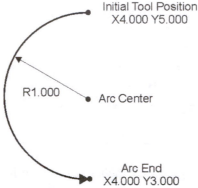

Initial Tool Position
X4.000 Y5.000

R1.000

Arc Center

Arc End
X4.000 Y3.000

20. Write a line of code using the incremental arc center method to machine the following arc at a feed rate of 7.5 inches per minute.

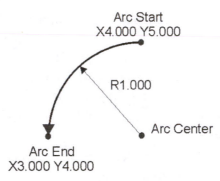

Arc Start
X4.000 Y5.000

R1.000

Arc Center

Arc End
X3.000 Y4.000

21. Write a line of code using the incremental arc center method to machine the following arc at a feed rate of 7 inches per minute.

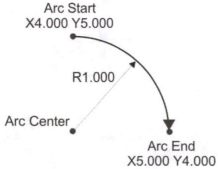

Arc Start
X4.000 Y5.000

R1.000

Arc Center

Arc End
X5.000 Y4.000

22. Write a line of code using the radius method to machine the following arc at a feed rate of 5 inches per minute.

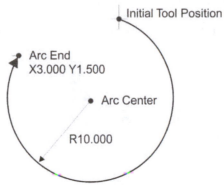

23. Write a line of code using the radius method to machine the following arc at a feed rate of 5 inches per minute.

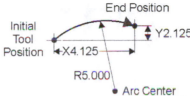

24. Write a line of code using the radius method to machine the following arc at a feed rate of 5 inches per minute.

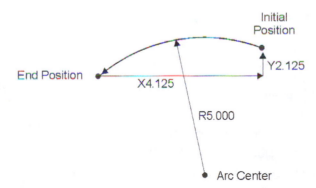

25. Write a line of code using the radius method to machine the following arc at a feed rate of 5 inches per minute.

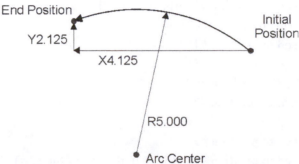

26. What purpose do tool length offsets serve?

27. What must be done to invoke tool diameter compensation?

28. Complete the blocks for the parts shown in the figure. Use a .50 end mill and cut to a depth of .25 around the part.

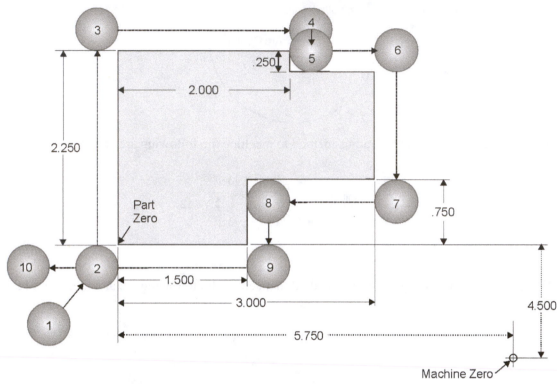

N10 G_ G_ G_ G_ G_; (Inch programming, absolute programming, cancel diameter compensation, cancel canned cycles, XY plane)
N20 M_ T02; (Tool change, tool #2)
N30 G54 X___ Y___; (Workpiece zero setting)
N40 M_ S800; (Spindle start clockwise, 800 RPM)
N50 G_ X-1.00 Y-1.00; (Rapid to position #1)
N60 G_ Z_; (Rapid down to .100 clearance above the part)
N70 G__ Z____ .5.0; (Feed down to depth at 5 inches per minute)
N80 G01 X__ Y___; (Feed to position #2, offsetting for the tool radius)
N90 G01 X___ Y___; (Feed to position #3)
N100 G01 X__ Y__; (Feed to position #4)
N110 G01 X__ Y__; (Feed to position #5)
N120 G01 X__ Y__; (Feed to position #6)
N130 G01 X__ Y__; (Feed to position #7)
N140 G01 X__ Y__; (Feed to position #8)
N150 G01 X__ Y__; (Feed to position #9)
N160 G01 X__ Y__; (Feed to position #10, 1 inch to the left of the part)

N170 G__; (Return all axes to home position)
N180 M__ T__; (Tool change, tool 0)
N190 M__; (End program, rewind program to beginning)

29. Program the part shown in the figure below. Use a .25 end mill to machine the outside shape of the part and machine to a depth of .35. Assume it is tool number 5. Make sure you use offsets. Program the part using the climb milling technique.

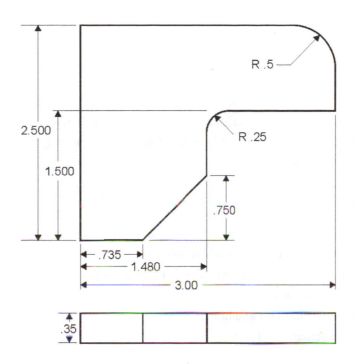

Chapter 6

PROGRAMMING EXAMPLES

This chapter will examine sample programs in a line-by-line fashion. The first program will be thoroughly explained. After the first program the rest will have explanations for the program commands that are new or might be difficult to understand.

OBJECTIVES

Upon completion of this chapter, the reader will be able to:
- Explain simple computer numerical control (CNC) programs line-by-line.
- Explain various G- and M-codes.
- Use codes to produce tool diameter and height offsets.
- Use codes to produce circular interpolation.
- Write CNC programs.

Programming Examples

The first program we examine will drill several 1/4-inch holes in a part. This example will use basic G-codes to drill the parts. Canned drilling cycles will be covered later. There are two example programs for this part. The first program uses absolute mode programming. The second uses incremental programming. Both programs use several G- and M-codes. Tool length offsets will also be used. Figure 6–1 shows the part.

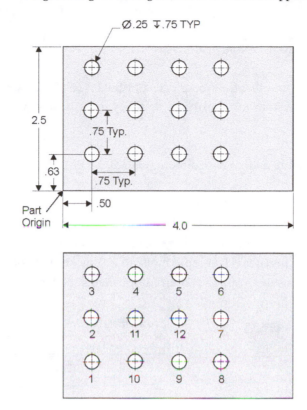

FIGURE 6–1. The part print is shown on top and the order of drilling on the bottom.

Programming can be thought of as a three-step process. We will follow a three-step process for these examples.

Step 1 contains the preparatory functions to get the control into the proper control modes.

Step 2 will contain the actual machining operations. There may be several. For example, step 2a might be a milling operation, step 2b might be a drilling operation, step 2c might be a tapping operation, and so on.

Step 3 ends the program correctly after all machining operations have been completed.

Program 1: Absolute Mode Programming

The first step in a program sets the control in the desired modes. Lines N10 through N20 are preparatory lines of code to get the control ready to run the part. Lines 10 and 20 set the desired modes for the control. The order of these codes at the beginning of a program does not matter.

O001 (Program name O001)

N10 G20

    Line N10 tells the control to operate in inch programming mode, not metric.

N20 G17 G40 G49 G80 G90

    G17 tells the control to operate in the XY plane. G40 cancels diameter offsets that may have been in the control from a previous program. G49 cancels tool length compensation that may have been in the control from a previous program. G80 cancels canned cycles that may have been in the control from a previous program. G90 tells the control to operate in absolute mode.

The second part of a program performs operations. This part of the program drills 12 holes. Only one tool is used. Tool 1 is a 1/4-inch drill.

N30 T1 M6

First a drill change loads tool 1 (Line N30).

N40 G54

In line N40 the control is told that the workpiece coordinates (X, Y, Z) are located in the G54 register. Note that the operator must make sure that they measure and put the correct values in the G54 register.

N50 S3500 M3

Then the spindle is turned in a clockwise direction at 3500 RPM. (Line N50).

N60 G00 X.5 Y.63

Next, a rapid move positions the tool over the location for the first hole -X.5 Y.630 (Line N60).

N70 G43 H1 Z.1 (Tool length compensation positive amount found in H1, rapid to Z.1)

In line N70, positive tool length compensation is called (see Figure 6-2). The amount of the compensation is found in register H1. A rapid move to Z.1 inches then occurs because the G00 is modal and still active from line N60. The control implements the length offset during the move to Z.1.

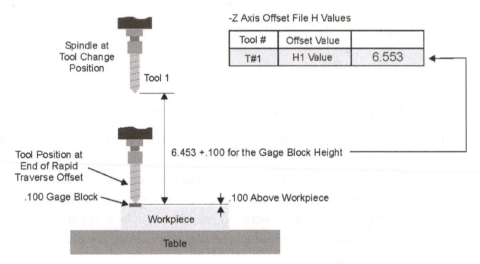

Figure 6-2. Tool length compensation amount for T1 (drill).

N80 G01 Z-.75 F8.0 (Linear move to Z-.75 at a feed rate of 8 inches per minute)

In line N80, the hole is drilled as the tool is fed to a Z depth of -.75 inches at a feed rate of 8.0 inches per minute.

N90 G00 Z.1 (Rapid move up to Z.1)

In line 90, the drill is moved with a rapid feed to Z.1 inches.

The rest of the program repeats these actions. A rapid feed positions the tool to the next hole location, the drill is fed to a depth of -.85 inches, the drill is moved with a rapid feed, back to Z.1 inches, and the process is repeated for the next hole.

The whole program is shown next.

O001 (Program name O001)
N10 G20 (Tells the control to operate in inch programming mode, not metric)
N20 G17 G40 G49 G80 G90

N30 T1 M6 (Change to tool 1, 1/4 drill)
N40 G54 (Workpiece coordinates are found in G54 register)
N50 S3500 M3 (Turn on spindle CW at a speed of 3500 RPM)
N60 G00 X.5 Y.63 (Rapid move to X.5 Y.630, hole 1)
N70 G43 H1 Z.1 (Tool length compensation positive amount found in H1, rapid to Z.1)
N80 G01 Z-.75 F8.0 (Linear move to Z-.75 at a feed rate of 8 inches per minute)
N90 G00 Z.1 (Rapid move up to Z.1)
N100 Y1.38 (Rapid move to Y 1.38, hole 2)
N110 G01 Z-.75 F8.0 (Linear move to Z-.75 at a feed rate of 8 inches per minute)
N120 G00 Z.1 (Rapid move up to Z.1)
N130 Y2.13 (Rapid move to Y2.13, hole 3)
N140 G01 Z-.75 F8.0 (Linear move to Z-.75 at a feed rate of 8 inches per minute)
N150 G00 Z.1 (Rapid move up to Z.1)
N160 X1.25 (Rapid move to X1.25, hole 4)
N170 G01 Z-.75 F8.0 (Linear move to Z-.75 at a feed rate of 8 inches per minute)
N180 G00 Z.1 (Rapid move up to Z.1)
N190 X2. (Rapid move to X2, hole 5)
N200 G01 Z-.75 F8.0 (Linear move to Z-.75 at a feed rate of 8 inches per minute)
N210 G00 Z.1 (Rapid move up to Z.1)
N220 X2.75 (Rapid move to X2.75, hole 6)
N230 G01 Z-.75 F8.0 (Linear move to Z-.75 at a feed rate of 8 inches per minute)
N240 G00 Z.1 (Rapid move up to Z.1)
N250 Y1.38 (Rapid move to Y1.38, hole 7)
N260 G01 Z-.75 F8.0 (Linear move to Z-.75 at a feed rate of 8 inches per minute)
N270 G00 Z.1 (Rapid move up to Z.1)
N280 Y.63 (Rapid move to Y.63, hole 8)
N290 G01 Z-.75 F8.0 (Linear move to Z-.75 at a feed rate of 8 inches per minute)
N300 G00 Z.1 (Rapid move up to Z.1)
N310 X2. (Rapid move to X2, hole 9)
N320 G01 Z-.75 F8.0 (Linear move to Z-.75 at a feed rate of 8 inches per minute)
N330 G00 Z.1 (Rapid move up to Z.1)
N340 X1.25 (Rapid move to Y1.25, hole 10)
N350 G01 Z-.75 F8.0 (Linear move to Z-.75 at a feed rate of 8 inches per minute)
N360 G00 Z.1 (Rapid move up to Z.1)
N370 Y1.38 (Rapid move to Y1.38, hole 11)
N380 G01 Z-.75 F8.0 (Linear move to Z-.75 at a feed rate of 8 inches per minute)
N390 G00 Z.1 (Rapid move up to Z.1)
N400 X2. (Rapid move to X2, hole 12)
N410 G01 Z-.75 F8.0 (Linear move to Z-.75 at a feed rate of 8 inches per minute)
N420 G00 Z.1 (Rapid move up to Z.1)

All machining is done at this point, and the program is ended in lines N430-N460.

N430 M5 (Stop spindle rotation)
N450 G28 (Return to reference point)
N460 M30 (Program end, memory reset)
%

Remember to think of a program as a three-step process.

Step 1: Get the control into the desired modes through the use of preparatory functions.

Step 2a ... 2n: Perform the machining operations.

Step 3: End the program.

Program 2: Incremental Mode Programming

The second example uses the same part (see Figure 6–3). This program will use incremental mode, however.

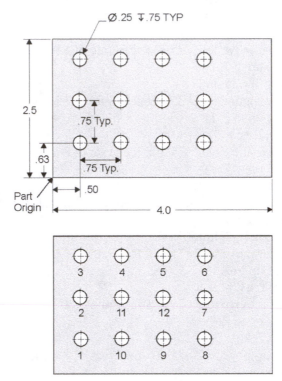

FIGURE 6–3. The part print is shown on the top and the drill order on the bottom.

Lines N10 through N20 are preparatory lines of code to get the control ready to run the part. Lines 10 and 20 set the desired modes for the control. Remember that the order of these preparatory codes, at the beginning of a program, does not matter.

O002 (Program name O002)

N10 G20 (Inch programming)

N20 G17 G40 G49 G80 G90 (Rapid, XY Plane, cancel diameter offsets, cancel tool length compensation, cancel canned cycles, absolute mode)

The second part of a program performs operations. This part of the program drills 12 holes. Only one tool is used. Tool 1 is a 1/4-inch drill. First a drill change loads tool 1 in line N30. Then the control is told that the workpiece coordinates (X, Y, Z) are located in the G54 register (line N40). Note that the operator must make sure that they measure and put the correct values in the G54 register. The spindle is turned on in Line N50.

In line N60 a rapid move positions the tool to the first hole location. In line N70, positive tool length compensation is called. The amount of the compensation is found in register H1. A rapid move to Z.1 then occurs. The control implements the length offset during the move to Z.1.

Line N80 puts the control into incremental mode. In line N90, the hole is drilled as the tool is fed to a Z depth of -.75 at a feed rate of 8.0 inches per minute. In line 100, the drill is moved with a rapid feed to Z.1. The rest of the program repeats these actions. A rapid positions the tool to the next hole location, the drill is fed to a depth of -.85, the drill is moved with a rapid back to Z.1, and the process is repeated for the next hole.

```
N30 T1 M6 (Change to tool 1, 1/4 drill)
N40 G54 (Workpiece coordinates are found in G54 register)
N50 S3500 M3 (Turn spindle on CW at 3500 RPM)
N60 G00 X.5 Y.63 (Rapid move to X.5 Y.63, hole 1)
N70 G43 H1 Z.1 (Positive tool height offset, value of offset in D1, move to Z.1)
N80 G91 (Incremental mode)
N90 G01 Z-.85 F8.0 (Linear move to Z-.85 at a feed rate of 8 inches per minute)
N100 G00 Z.85 (Rapid .85 in the Z+ direction)
N110 Y.75 (Rapid move .75 in the Y+ direction, hole 2)
N120 G01 Z-.85 F8.0 (Linear move to -.85 in the Z- direction at 8 inches per minute)
N130 G00 Z.85 (Rapid .85 in the Z+ direction)
N140 Y.75 (Rapid move .75 in the Y+ direction, hole 3)
N150 G01 Z-.85 F8.0 (Linear move to -.85 in the Z- direction at 8 inches per minute)
N160 G00 Z.85 (Rapid .85 in the Z+ direction)
N170 X.75 (Rapid move .75 in the X+ direction, hole 4)
N180 G01 Z-.85 F8.0 (Linear move to -.85 in the Z- direction at 8 inches per minute)
N190 G00 Z.85 (Rapid .85 in the Z+ direction)
N200 X.75 (Rapid move .75 in the X+ direction, hole 5)
N210 G01 Z-.85 F8.0 (Linear move to -.85 in the Z- direction at 8 inches per minute)
N220 G00 Z.85 (Rapid .85 in the Z+ direction)
N230 X.75 (Rapid move .75 in the X+ direction, hole 6)
N240 G01 Z-.85 F8.0 (Linear move to -.85 in the Z- direction at 8 inches per minute)
N250 G00 Z.85 (Rapid .85 in the Z+ direction)
N260 Y-.75 (Rapid -.75 in the Y- direction, hole 7)
N270 G01 Z-.85 F8.0 (Linear move to -.85 in the Z- direction at 8 inches per minute)
N280 G00 Z.85 (Rapid .85 in the Z+ direction)
N290 Y-.75 (Rapid move to -.75 in the Y- direction, hole 8)
N300 G01 Z-.85 F8.0 (Linear move to -.85 in the Z- direction at 8 inches per minute)
N310 G00 Z.85 (Rapid .85 in the Z+ direction)
N320 X-.75 (Rapid move to -.75 in the X- direction, hole 9)
N330 G01 Z-.85 F8.0 (Linear move to -.85 in the Z- direction at 8 inches per minute)
N340 G00 Z.85 (Rapid .85 in the Z+ direction)
N350 X-.75 (Rapid move to -.75 in the X- direction, hole 10)
```

N360 G01 Z-.85 F8.0 (Linear move to -.85 in the Z- direction at 8 inches per minute)

N370 G00 Z.85 (Rapid .85 in the Z+ direction)

N380 Y.75 (Linear move to .75 in the Y+ direction, hole 11)

N390 G01 Z-.85 F8.0 (Linear move to -.85 in the Z- direction at 8 inches per minute)

N400 G00 Z.85 (Rapid .85 in the Z+ direction)

N410 X.75 (Rapid move .75 in the X+ direction, hole 12)

N420 G01 Z-.85 F8.0 (Linear move to -.85 in the Z- direction at 8 inches per minute)

N430 G00 Z.85 (Rapid .85 in the Z+ direction)

All machining is done at this point, and the program is ended in lines N440-N460

N440 M5 (Spindle stop)

N450 G28 (Return to reference point)

N460 M30 (Program end, memory reset)

%

Program 3: Milling a Groove

This program mills a triangular groove in a part. No tool diameter offsets will be used in this program. Positive-tool height compensation will be used. The part is shown in Figure 6–4.

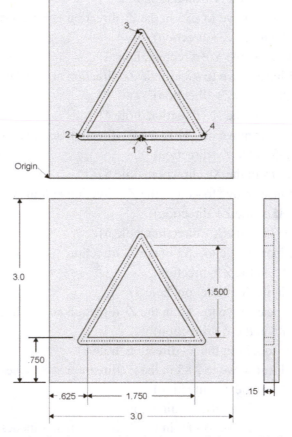

FIGURE 6–4. The part print is shown on the bottom and the machining path on the top. The part will be machined with a 1/4-inch end mill.

O003 (Program name O003)

Lines N10 through N20 are preparatory lines of code to get the control ready to run the part. Lines 10 and 20 set the desired modes for the control. Remember that the order of these at the beginning of a program does not matter.

N10 G20 (Inch mode)

N20 G17 G40 G49 G80 G90 (Rapid, XY Plane, cancel diameter offsets, cancel tool length compensation, cancel canned cycles, absolute mode)

Line N30 tells the control that the workpiece coordinates (X, Y, Z) are located in the G54 register. Note that the operator must make sure that they put the correct values in the G54 register.

N30 G54 (Workpiece coordinates are found in G54 register)

N40 T1 M6 (Change to tool 1)

N50 S5000 M3 (Turn on spindle clockwise at a speed of 5000 RPM)

N60 G00 X1.5 Y.625 (Linear move to X1.5 Y.625, point 1)

Line N70 calls for positive tool length compensation by the amount in H1 and rapids the tool to .1 above the workpiece. The control implements the length offset during the move to Z.1.

N70 G43 H1 Z1.

N80 G00 Z.1 (Rapid to Z.1)

N90 G01 Z-.15 F6. (Linear move to Z-.15 at a feed rate of 6.0 inches per minute)

N100 G01 X.4074 F10. (Linear move to X1.5 Y.625 at a feed rate of 10.0 inches per minute, point 2)

N110 G01 X1.5 Y2.4981 (Linear move to X1.5 Y2.4981, point 3)

N120 G01 X2.5926 Y.625 (Linear move to X2.5926 Y.625, point 4)

N130 G01 X1.5 (Linear move to X1.5, point 5)

N140 G01 Z.1 F6. (Linear move to Z.1 at a feed rate of 6.0 inches per minute)

N150 G0 Z1. (Rapid to Z.1)

The rest of the lines stop the spindle, return to reference, and end the program.

N160 M5 (Spindle stop)

N170 G28 (Return to reference point)

N180 M30 (Program end, memory reset)

%

Program 4: Milling With Circular Interpolation

This part will involve milling a groove in a part that involves circular interpolation. The part is shown in Figure 6–5. The cutter will be a 1/4-inch end mill.

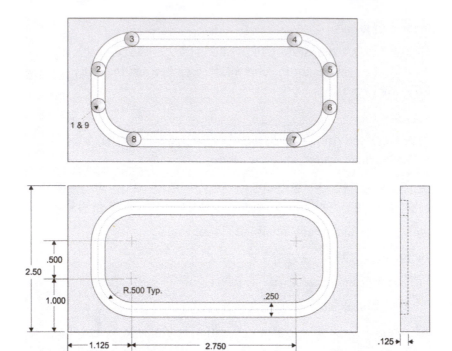

FIGURE 6–5. The part print is shown on the bottom and the machining path order on the top.

Lines N10 and N20 are the preparatory codes to set the correct modes for the control. Line N30 tells the control that the coordinates for the workpiece are in register G54. Note that the operator must make sure that they measure and put the correct values in the G54 register.

O004 (Program name O004)
N10 G20 (Inch mode)
N20 G17 G40 G49 G80 G90 (Rapid, XY Plane, cancel diameter offsets, cancel tool length compensation, cancel canned cycles, absolute mode)
N30 G54 (Workpiece coordinates are found in G54 register)
N40 T1 M6 (Load tool 1, 1/4 end mill)
N50 S5000 M3 (Turn on spindle clockwise at a speed of 5000 RPM)
N60 G00 X.63 Y1. (Rapid move to X.63 Y1, point 1)
    Line N70 calls for positive tool length compensation by the amount in H1 and rapids the tool to 1.0. inches above the workpiece. The control implements the length offset during the move to Z.1.
N70 G43 H1 Z1. (Positive tool height offset, value of offset in D1, move to Z.1)
N80 G01 Z.1 (Linear move to Z.1)
N90 G01 Z-.13 F6.0 (Linear move to Z-.13 at a feed rate of 6.0 inches per minute)
    Line N100 is a clockwise circular interpolation to point 3. The X1. and the Y2. are the end point of the circular move. The I value is the value in the X direction between the start point and the center of the arc. An I value of .5 means that the arc center is .5 inches to the right (the plus direction) the start point in the X direction. The J value is not listed, which means that the arc center location in the Y direction is the same as the start point Y location value.
N100 G01 Y1.5 F20. (Linear move to point 2 at a feed rate of 20 inches per minute)
    Line N100 is a clockwise circular interpolation to point 3. The X1. and the Y2. are the end point of the circular move.

The I value is the value in the X direction between the start point and the center of the arc. An I value of .5 means that the arc center is .5 inches to the right (the plus direction) the start point in the X direction. The J value is not listed, which means that the arc center location in the Y direction is the same as the start point Y location value.

N110 G02 X1.13 Y2. I.5 (Clockwise circular interpolation to point 3)

N120 G01 X3.88 (point 4)

Line N130 is a clockwise circular interpolation to point 5. The X4.38 and the Y1.5 are the end point of the circular move. The I value is not listed, which means that the arc center in the X direction is the same as the start point X value. The J value is the value in the Y direction between the start point and the center of the arc. A J value of -.5 means that the arc center is -.5 below (the minus direction) the start point in the Y direction.

N130 G02 X4.38 Y1.5 J-.5 (Clockwise circular interpolation to point 5)

N140 G01 Y1 (point 6)

Line N150 is a clockwise circular interpolation to point 7. The X3.88 and the Y.5 are the end point of the circular move. The I value is the value in the X direction between the start point and the center of the arc. A I value of -.5 means that the arc center is -.5 inch to the left (the minus direction) the start point in the X direction. The J value is not listed, which means that the arc center location in the Y direction is the same as the start point Y location value.

N150 G02 X3.88 Y.5 I-.5 (Clockwise circular interpolation to point 7)

N160 G01 X1.13 (Linear move to point 8)

Line N170 is a clockwise circular interpolation to point 9. The X.63 and the Y1. are the end point of the circular move. The I value is not listed, which means that the arc center in the X direction is the same as the start point X value. The J value is the value in the Y direction between the start point and the center of the arc. A J value of .5 means that the arc center is .5 above (the plus direction) the start point in the Y direction.

N170 G02 X.63 Y1. J.5 (Clockwise circular interpolation to point 9)

N180 G01 Z.1 F6.0 (Linear move to Z.1 at a feed rate of 6. inches per minute)

N190 G00 Z1. (Rapid to Z1.)

The next lines stop the spindle, return to the reference position, and end the program.

N200 M5 (Spindle stop)

N210 G28 (Return to reference point)

N220 M30 (Program end, memory reset)

%

Program Example 5: Circular Interpolation Using the IJ Programming Method

The next part will involve tool compensation and circular interpolation. This program will use the IJ method of circular interpolation. The part is shown in Figure 6–6.

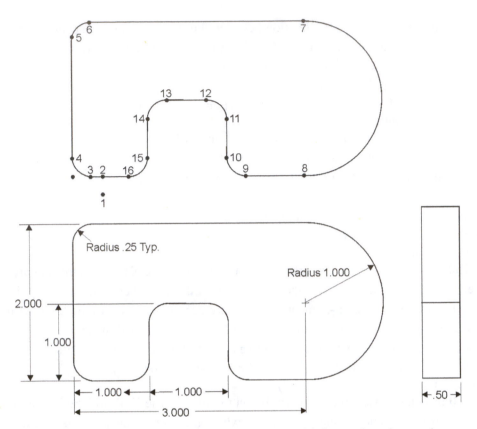

FIGURE 6–6. The part print is shown on the bottom and the machining path order on the top.

O005 (Program name O005)
   Lines N10 through N20 are the preparatory lines. Line N10 sets the desired modes for the controller.
N10 G17 G20 G40 G80 G90 (XY Plane, inch mode, cancel diameter offsets, cancel canned cycles, absolute mode)
N20 G54 (Workpiece coordinates are found in G54 register)
N30 T1 M06 (Change to tool 1, .375 end mill)
N40 S4000 M3 (Turn on spindle clockwise at a speed of 4000 RPM)
N50 G00 X.375 Y-.375 (Rapid move to X.375 Y-.375, point 1)
N60 G43 H1 Z.1 (Positive tool length compensation by the amount in H1 and rapids the tool to .1 above the workpiece. The compensation amount is implemented during the move to Z.1)
N70 G01 Z-.5 F6.0 (Feeds the tool to a depth of Z-.5 at a feed rate of 6. inches per minute)

Line N80 tells the control to offset the tool to the left of the cutter path by the amount found in D1. In this example, the cutter is .375 diameter so the offset would be 1/2 of that size (.1875). The operator should make sure that D1 contains .1875. Line N80 also performs a linear move to Y0. at a feed rate of 10.0 inches per minute. During this move, the control implements the offset. When the cutter arrives at Y0.0, the control has offset the cutter to the left by .1875.

   N80 G41 D1 Y0. F10.0 (Tool diameter compensation left by the amount in D1, linear move to Y0. at a feed rate of 10.0 inches per minute, point 2)

Line N90 moves to point 3. Line N100 is a clockwise circular interpolation to point 4. The X0. and the Y.25 are the end point of the circular move. The I value is not listed, which means that the arc center in the X direction is the same as the start point X value. The J value is the value in the Y direction between the start point and the center of the arc. A J value of .25 means that the arc center is .25 above (the plus direction) the start point in the Y direction.

```
N90 G01 X.25 (Linear move to X.25, point 3)
N100 G02 X0. Y.25 J.25 (Clockwise circular interpolation, point 4)
N110 G01 Y1.75 (Linear move to Y1.75, point 5)
```

Line N120 is another clockwise circular interpolation. The end point of the move is at X.25 Y2. The I value of .25 means that the arc center in the X direction is .25 from the start point.

```
N120 G02 X.25 Y2. I.25 (Clockwise circular interpolation, point 6)
N130 G01 X3. (Linear move to X3., point 7)
N140 G02 X3. Y0. J-1. (Clockwise circular interpolation, point 8)
N150 G01 X2.25 (Linear move to X2.25, point 9)
N160 G02 X2. Y.25 J.25 (Clockwise circular interpolation, point 10)
N170 G01 Y.75 (Linear move to Y.75, point 11)
N180 G03 X1.75 Y1. I-.25 (Counterclockwise circular move, point 12)
N190 G01 X1.25 (Linear move to X1.25, point 13)
N200 G03 X1. Y.75 J-.25 (Counterclockwise circular move, point 14)
N210 G01 Y.25 (Linear move to Y.25, point 15)
N220 G02 X.75 Y0. I-.25 (Clockwise circular interpolation, point 16)
N240 G01 X-.375 (Linear move to X-.375, point 17)
N250 G01 Y-.375 (Linear move to Y.375, point 18)
```

Note that line N260 cancels the diameter offset that was called for with this tool.

```
N260 G40
N270 G01 Z.1 F6.0
```

Lines N280 through the end correctly shut down the machine. Line N280 stops the spindle. Line N290 sets the control in incremental mode and returns the machine to the reference position.

```
N280 G00 M5 (Rapid mode, spindle stop)
N290 G28 (Return to reference point)
N300 M30 (Program end, memory reset)
%
```

Program 6: Circular Interpolation Using the Radius Programming Method

The next part will involve tool compensation and circular interpolation. This program will use the radius method of circular interpolation. The part is shown in Figure 6–7.

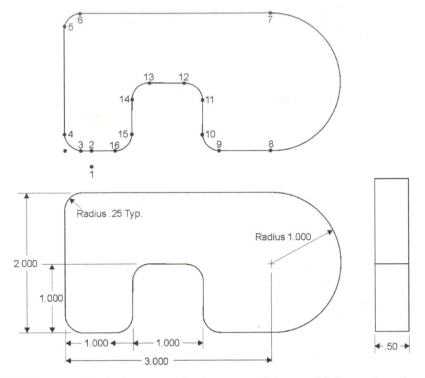

FIGURE 6–7. The part print is shown on the bottom and the machining path order on the top.

Lines N10 through N20 are the preparatory lines. Line N10 sets the desired modes for the controller. Line N20 sets the workpiece coordinates.

O006 (Program name O006)

N10 G17 G20 G40 G80 G90 (Rapid, XY Plane, cancel diameter offsets, cancel tool length compensation, cancel canned cycles, absolute mode)

N20 G54 (Workpiece coordinates are found in G54 register)

N30 T1 M06 (Tool Change to tool 1, .375 END MILL)

N40 S4000 M3 (Turn on spindle clockwise at a speed of 4000 RPM)

N50 G00 X.375 Y-.375 (Rapid move to X.375 Y-.375, point 1)

Line N60 invokes positive tool length compensation by the amount in register H1. The control implements the compensation in this line as it moves the cutter to Z.1.

N60 G43 H1 Z.1

N70 G01 Z-.5 F6.0 (Linear move to Z-.5 at a feed rate of 6.0 inches per minute, point)

Line N80 tells the control to offset the tool to the left of the cutter path by the amount found in D1. In this example, the cutter is .375 diameter, so the offset would be 1/2 of that size (.1875). The operator should make sure that D1 contains .1875. Line N80 also performs a linear move to Y0. at a feed rate of 15.0 inches per minute. During this move, the control implements the offset. When the cutter arrives at Y0., the control has offset the cutter to the left by .1875.

N80 G41 D1 Y0. F15.0 (Linear move to Y0., point 2)
N90 G01 X.25 (Linear move to X.25, point 3)

Line N100 is a clockwise circular interpolation. The end point for the move will be at X0. Y.25. The r value of .25 is the radius of the arc.

N100 G02 X0. Y.25 R.25 (Clockwise circular move to point 4)
N110 G01 Y1.75 (Linear move to Y1.75, point 5)

Line N120 is a clockwise circular interpolation. The end point for the move will be at X.25 Y2. The r value of .25 is the radius of the arc.

N120 G02 X.25 Y2. R.25 (Clockwise circular move to point 6)
N130 G01 X3. (Linear move to X3., point 7)
N140 G02 X4. Y1. R1. (Clockwise circular move to point 8)
N150 G02 X3. Y0. R1. (Clockwise circular move to point 9)
N160 G01 X2.25 (Linear move to X2.25, point 10)
N170 G02 X2. Y.25 R.25 (Clockwise circular move to point 11)
N180 G01 Y.75 (Linear move to Y.75, point 12)
N190 G03 X1.75 Y1. R.25 (Counterclockwise circular move, point 13)
N200 G01 X1.25 (Linear move to Y1.25, point 14)
N210 G03 X1. Y.75 R.25 (Counterclockwise circular move, point 15)
N220 G01 Y.25 (Linear move to Y.25, point 16)
N230 G02 X.75 Y0. R.25 (Clockwise circular move to point 17)
N240 G01 X-.375 (Linear move to X-.375, point 18)
N250 G01 Y-.375 (Linear move to Y-.375, point 19)

Note that line N260 cancels the diameter offset that was called for this tool.

N260 G40 (Cancel tool diameter compensation)

N270 G01 Z.1 F6.0 (Linear move to Z.1 at a feed rate of 6.0 inches per minute)

Lines N280 through N310 stop the spindle, return to reference position, and perform a program end and memory reset.

N280 M5 (Stop spindle)
N290 G28 (Return to reference point)
N300 M30 (Program end, memory reset)
%

Program 7: Milling Example

The next program will be another milling example. The part is shown in Figure 6–8. It will use height and diameter compensation. There is nothing new introduced in this program. The only explanation for this program will be in each line of code.

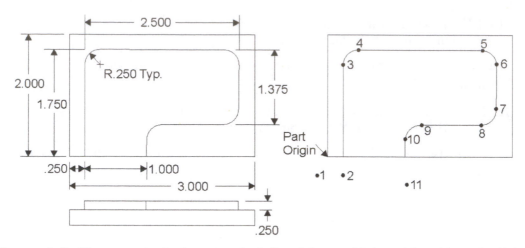

FIGURE 6–8. The part print is shown on the left and the machining path order on the right.

O007 (Program name O007- Job)
N10 G20 (inch mode)
N20 G17 G40 G49 G80 G90 (Rapid, XY Plane, cancel diameter offsets, cancel tool length compensation, cancel canned cycles, absolute mode)
N30 G54 (Workpiece coordinates are found in G54 register)
N40 T1 M6 (3/8 FLAT ENDMILL TOOL)
N50 S1400 M3 (Turn on spindle clockwise at a speed of 1400 RPM)
N60 G00 X-.25 Y-.25 (Rapid move to X-.25 Y-.25, point 1)

Line N70 calls for positive tool length compensation by the amount in H1 and rapids the tool to .1 above the workpiece. The compensation amount is implemented during the move to Z.1

N70 G43 H1 Z1. (G43 H1- Tool height offset amount in H3, rapid to Z1.)
N80 Z.1 (Rapid to Z.1)
N90 G01 Z-.25 F6.00 (Linear feed to Z-.25 at a feed rate of 6.00 inches per minute)
N100 G41 D1 X.25 (G41- cutter compensation left by amount in D1, linear move to X.25, point 2)
N110 G01 Y1.5 (Linear move to Y1.5, point 3)
N120 G02 X.5 Y1.75 R.25 (Clockwise circular interpolation, point 4)
N130 G01 X2.5 (Linear move to Y1.5, point 5)
N140 G02 X2.75 Y1.5 R.25 (Clockwise circular interpolation, point 6)
N150 G01 Y.62 (Linear move to Y1.5, point 7)
N160 G02 X2.5 Y.37 R.25 (Clockwise circular interpolation, point 8)
N170 G01 X1.5 (Linear move to Y1.5, point 9)
N180 G03 X1.25 Y.12 R.25 (Clockwise circular interpolation, point 10)
N190 G01 Y-.50 (Linear move to Y1.5, point 11)

Note that line N200 cancels the diameter offset that was called for this tool.

N200 G40 Z.1 (G40 Cancel tool diameter compensation, move to Z.1)
N210 G00 Z1. (Rapid move to Z1.)
N220 M5 (Spindle stop)
N230 G28 (Return to reference point)
N240 M30 (Program end, memory reset)
%

Program 8: Height and Diameter Compensation Example

The next program will be another milling example. The part is shown in Figure 6–9. It will use height and diameter compensation. There is nothing new introduced in this program. The only explanation for this program will be in each line of code.

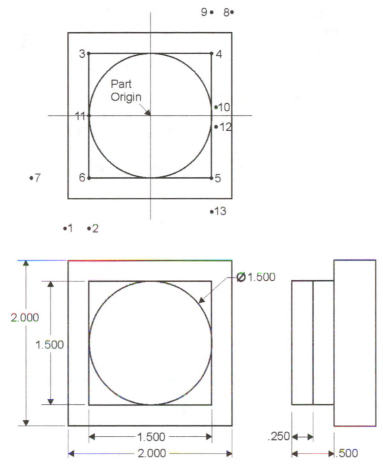

FIGURE 6–9. Milling example.

O008 (Program name O008)
N10 G20 (Inch mode)
N20 G17 G40 G49 G80 G90 (Rapid, XY Plane, cancel diameter offsets, cancel tool length compensation, cancel canned cycles, absolute mode)
N30 G54 (Workpiece coordinates are found in G54 register)
N40 T1 M6 (Load tool 1, 3/4 end mill)
N50 S2000 M3 (Turn on spindle clockwise at a speed of 2000 RPM)
N60 G00 X-1.25 Y-1.75 (Rapid move to X-1.25 Y-1.75., point 1)
N70 G43 H1 Z1. (G43 H1 – Positive tool length compensation amount in H1, rapid to Z1.)
N80 G01 Z-.5 F6. (Linear feed to Z-.5 at a feed rate of 6.)
N90 G41 D1 X-.75 F24. (Cutter diameter compensation left amount in D1, linear move to X-.75, point 2)
N100 G01 Y.75 (Linear move to Y.75, point 3)

123

N110 G01 X.75 (Linear move to X.75, point 4)
N120 G01 Y-.75 (Linear move to Y-.75, point 5)
N130 G01 X-.75 (Linear move to X-.75, point 6)
N140 X-1.875 (Linear move to X-1.875, point 7)
N150 G01 G40 Z.1 F6. (Linear move to Y.75)
N160 G00 Z1. (Rapid to Z1.)
N170 G00 X1.00 Y1.125 (Rapid move to X1.00 Y1.125, point 8)
N180 G00 Z.1 (Rapid move to Z.1)
N190 G01 Z-.25 (Linear feed to Z-.25)
N200 G41 D1 X.75 (Left diameter cutter compensation by amount in D1, linear move to X.75, point 9)
N210 G01 Y0. F24. (Linear move at 24. inches per minute, point 10)
N210 G02 X-.75 Y0. I-.75 (Clockwise circular interpolation, point 11)
N220 G02 X.75 Y0. I.75 (Clockwise circular interpolation, point 12)
N230 G01 Y-1.125 (Linear move to Y.75, point 13)
N240 G01 G40 Z.1 F6. (G40 – Cancel tool diameter compensation, G01 Z.1 F6. – Linear move to Z.1 at a feed rate of 6. inches per minute)
N250 G00 Z1. (Rapid to Z1.)
N260 M5 (Spindle stop)
N270 G28 (Return to reference point)
N280 M30 (Program end, memory reset)
%

Chapter Questions

1. Program the part shown in Figure 6–10. Use a .5 end mill to machine the outside shape of the part and machine to a depth of .35. Assume it is tool number 5. Make sure you use height and diameter offset compensation. Program the part using the conventional milling technique (counter-clockwise).

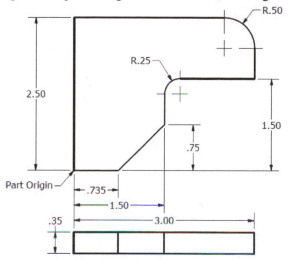

FIGURE 6–10. Use with question 1.

2. Program the part shown in Figure 6–11. Use a .5 end mill to machine the outside shape of the part and machine to a depth of .50. Assume it is tool number 5. Make sure you use height and diameter offset compensation.

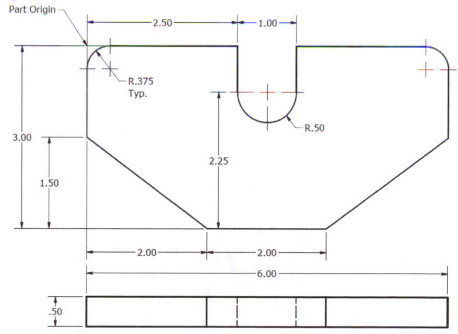

FIGURE 6–11. Use with question 2.

3. Program the part shown in Figure 6–12. Use a .5 inch end mill. Mill the contour to a .50 depth.

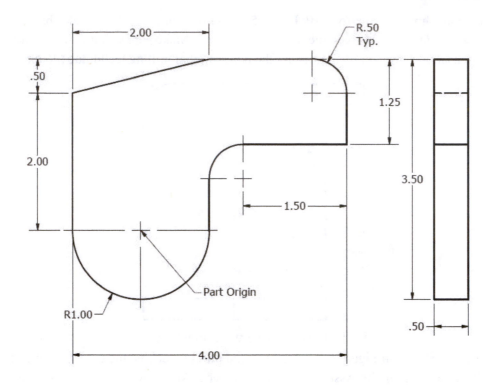

FIGURE 6–12. Programming exercise.

Chapter 7

PROGRAMMING CANNED CYCLES FOR MACHINING CENTERS

INTRODUCTION

This chapter will cover the steps necessary to properly plan, set up, and program a machining center. This chapter will focus on canned cycles. Canned cycles are sometimes called fixed cycles. Canned cycles will reduce the amount of programming necessary for repetitive machining operations such as drilling, boring, reaming, and tapping.

OBJECTIVES

Upon completion of this chapter, the reader will be able to:
- Demonstrate an understanding of acceptable machining center programming practices.
- Describe the typical sequence of operations for machining centers.
- Describe the steps necessary to properly plan a program.
- State the purpose of a setup sheet.
- Program a machining center using linear moves, circular moves, and canned cycles.

Planning the Program

The first step in preparing any program is to carefully plan the steps needed to make the part. This preparation is called a *process plan*. The process plan outlines all of the steps and tooling needed to machine the part. The first step in planning the program is to study the part drawing carefully.

The Part Drawing

The part drawing provides detailed information about the part. The shape of the part, the tolerances, material requirements, surface finishes, and the quantity required all have an impact on the program. From the part drawing, the programmer will determine what type of machine is required, work-holding considerations, and part datum (workpiece zero) location. Once these questions have been answered, the programmer can develop a process plan.

Choosing the Machine

The machine will be selected based on its size, horsepower, accuracy, tooling capacity, and the number of axes of travel required. The size of the machine is based on the amount of travel of the axes.

Does the part safely fit on the machine? Machine tools have a weight capacity. It can be unsafe to overload the capacity of the machine. What are the machining power requirements? The size of tools, depth of cut, and part material have a direct effect on the horsepower requirements of the machine.

Is the machine rigid enough to withstand rough-machining operations? The rigidity of the machine affects the depth of cut, feed and speed rates, and surface finishes. To maximize production rates, choose a machine that is rigid enough to handle the task.

What are the part tolerances? Is our machine capable of the accuracy required? How many different tools will be needed to manufacture this part? Will the machine's tool carousel accept that many tools? Will the machine be available when you are ready to make these parts?

Work Holding

Will fixtures to hold the part be required? Fixtures need to be planned early in the process. This will ensure that the proper lead time is allotted for purchasing or manufacturing this type of holding device. Small square or rectangular parts can be held in a vise.

For large parts, a setup using strap clamps may be used. If the part is small but has a complicated shape, soft vise jaws may be machined to hold the part. Some typical methods of work holding were shown in Chapter 5.

Part Datum Location

The part datum location is usually based on how the part designer drew the part. When dimensioning a part, the designer typically uses a part feature that has a direct influence on how the part is used. If the designer dimensions the majority of the part features from the corner of the part, then we need to use that corner as the part datum. If the designer uses the center of a hole as the datum feature, the hole center should probably be used as the programming datum.

Selecting the Proper Tooling

Decide the tooling requirements for the part ahead of time. If special tools are required, make sure they are ordered and will arrive in time. When standard tools are going to be used, make sure you have enough of them on hand to complete the job. Waiting for tooling can be very costly.

Deciding on the types of tools to be used greatly influences your program. Carbide or specially coated tools can greatly speed up the machining process and should be used whenever possible.

Cost is the major factor in deciding on the tooling. Higher priced tooling may lower the total cost of machining because of decreased cycle time. When considering the cost of the tool, remember to take into account the cycle time to produce the part, the part tolerance, surface finish, and quantity of parts needed.

The Process Plan

Process planning involves deciding when certain machining operations will take place. Primary machining operations will take place on the CNC machine. The part configuration will normally determine the sequence of operations. There are some general rules to follow when deciding on the order of machining operations. The recommended procedures for machining are as follows:
- Face mill the top surface
- Rough machine the profile of the part
- Rough bore
- Drill and tap
- Finish profile surfaces
- Finish bore
- Finish reaming

Note that all roughing operations took place first: then finishing operations were done. This minimizes the effect of high-pressure operations moving or stressing the part.

Sometimes secondary operations may be done more economically on other types of machines. Secondary operations often include deburring the part after machining. Secondary operations may be done by the same operator while another part is running. This keeps the CNC machine making parts while the operator does the less demanding secondary operations on a simpler, less expensive machine. This approach to process planning is usually done by the operator/programmer in smaller job shop settings. In large shops, the process plan would come down from the engineering area and would include information for each step in the manufacturing of the part. In a small job-shop setting, the operator may act as the manufacturing engineer.

The operator must determine the best, most economical way to produce the part. The operator must determine the machining sequence, types of cutting tools, work-holding devices, and the machining conditions (cutting speed, feed rate, and depth of cut). Whether process planning is done by the manufacturing engineer or the operator, a plan for each operation should be developed. The process plan is done on a process planning sheet (see Figure 7–1).

Acme Machining Inc.- Process Plan			Part Datum		Part #
Operation	Tool #		Tool Description	RPM	Feed rate

FIGURE 7–1. The process plan is an outline of the machining steps to be done on the part.

The Setup Sheet

The CNC setup sheet is a detailed explanation of how the part will be set up, which fixture is to be used, where the part datum is located, and which tools are to be used (see Figure 7–2).

ACME Machining Inc. Setup Sheet		Part #
		Sheet /
Machine:		Prepared By:
Machining Operations		Setup Sketch
#	Description	
Comments:		

FIGURE 7–2. The CNC setup sheet may include a sketch or picture of the setup for clarity. Setup sheets help assure consistent quality, rapid setups, and consistent job setups and machining.

Setup sheets show operators what the programmer had in mind when he or she programmed the part. In small job shops, the programmer and the operator are usually the same person, but the setup sheet can be a useful review if the same or similar parts are run in the future. The setup sheet should contain all of the necessary information for the job and may include pictures of parts and fixtures. Setup sheets can dramatically reduce setup times.

Programming the Part

After all of the preliminary steps are done, it is time to write the program. If you have done a good job of planning the job, programming the part should be fairly simple and straightforward. While almost all programming is now done on the CNC or on a computer, a good way to learn is to manually write some programs. If you are writing the program by hand, a form can be used to keep the program organized (see Figure 7–3). Programming sheets keep track of the sequence of operation, coordinates, tool numbers, and miscellaneous functions.

ACME Machining Inc. - CNC Programming Sheet										
Part #				Programmer:					Date:	
N SEQ #	G Prep. Function	Axis Coordinates			Arc Center			F Feed	M Misc. Function	S RPM	H D L
		X	Y	Z	I	J	K				

FIGURE 7–3. Programming sheets can reduce errors and keep the program orderly.

Canned Cycles for Machining Centers

Canned cycles (fixed cycles) simplify the programming of repetitive machining operations such as drilling, tapping, and boring. Canned cycles are a set of preprogrammed instructions that eliminate the need for many lines of programming. Programming a simple drilled hole, without the use of a canned cycle, can take four or five lines of programming. Think of the lines that are needed to produce a hole:

- Position the X and Y axis to the proper coordinates with a rapid traverse move (G00),
- Position the Z axis to a clearance plane,
- Feed the tool down to depth,
- Rapid position the tool back to the clearance point.

That is 4 steps for one drilled hole! By using a canned drill cycle, a hole can be done with one line of code. Standard canned cycles, or fixed cycles, are common to most CNC machines. Figure 7–4 lists a few commonly used canned cycles for machining centers.

G-Code	Function	Z Axis	At Depth	Z Axis Return
G81	Drill	Feed	-	Rapid Traverse
G83	Peck Drill	Feed with Peck	-	Rapid Traverse
G84	Tapping	Feed	Reverse Spindle	Feed
G85	Bore/Ream	Feed	Stop Spindle	Cutting Feed
G86	Bore	Feed	Stop Spindle	Rapid Traverse

FIGURE 7–4. A few canned cycles for machining centers.

G81 Canned Drilling Cycle

One of the most commonly used canned cycle is the G81 canned drilling cycle. This cycle will automatically do all of the things necessary to drill a hole in one line of code (see Figure 7–5). The Z position is very important when you call the canned cycle. The present Z position will become the Z initial plane. The machine will normally rapid back to the Z initial position before a rapid to the next hole. If the tool is 12 inches above the work when the canned cycle is called, that will be the Z initial plane, and the machine will rapid up to 12 inches above the work between each hole. This would be a very inefficient program.

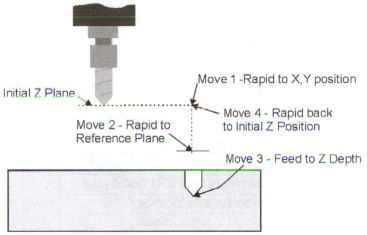

FIGURE 7–5. A G81 canned-drill cycle.

As you can see from Figure 7–5, the canned drilling cycle consists of four moves. Those four moves are controlled by one line of programming. Study Figure 7-6 for an example of a G81 canned cycle.

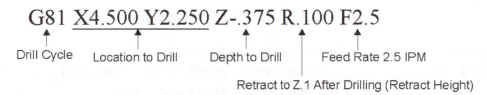

FIGURE 7–6. Example of a G81 canned drill cycle.

The G81 specifies a drill canned cycle, X and Y are the coordinates of the hole center, Z is the depth of the hole, R is the height for the drill to rapid down to from the initial Z position. F is the feed rate in inches per minute.

Figure 7-7 shows two examples of the use of a G81. The part on the left has a raised section that needs to be avoided on moves so the tool will not be broken. The part on the right has no obstruction. If we want to use the G81 drill cycle and retract up the initial Z we would use a G98 with it. A G98 means retract to the initial Z position after drilling. The part on the right has no obstruction on the top of the part. The drill cycle will be quicker if the drill only retracts to the R plane before the rapid move to the next hole location. A G99 would be used with the G81 to make the drill retract to the R plane instead of the initial Z position.

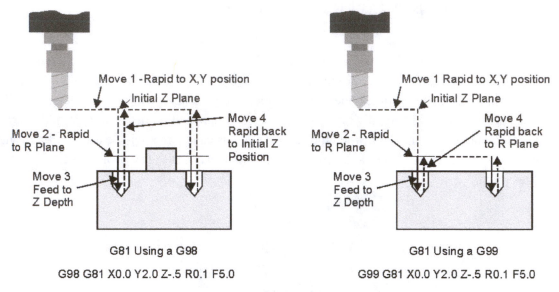

G81 Using a G98

G98 G81 X0.0 Y2.0 Z-.5 R0.1 F5.0

G81 Using a G99

G99 G81 X0.0 Y2.0 Z-.5 R0.1 F5.0

FIGURE 7–7. G81 canned cycle examples.

Figure 7-8 shows the parameters for a G81 drilling cycle.

Parameters for a G81 Drill Cycle	
X*	Rapid X axis location
Y*	Rapid Y axis location
Z	Depth of the hole
R	R Plane
F	Feed rate
*Optional	

Figure 7-8. Parameters for a G81 canned drilling cycle.

The G81 canned cycle is modal, which means that it will stay active until it is canceled by a G80. If we were drilling a series of holes, we would only need to specify the coordinates for the next hole. Figure 7–9 and the program that follows are a drilling example using a G81 canned drill cycle using a G98 code that makes the tool retract to the R plane after each drill cycle.

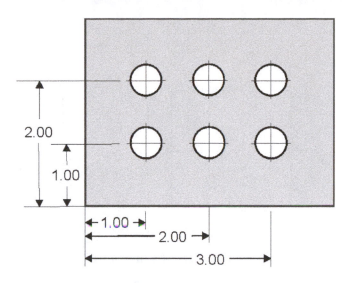

FIGURE 7–9. Canned drilling cycle example.

O00079 (Program name O00079)

N20 G00 G17 G40 G49 G80 G90 Rapid mode, G17 - XY plane, G40 - Cancel diameter compensation offset, G49 - Cancel tool length compensation, G90 - Absolute mode)

N30 T1 M06 (Tool Change to tool 1, 1/2 drill)

N40 G54 (Workpiece zero setting)

N50 S1000 M03 (1000 RPM, Spindle start clockwise)

N60 G00 X1. Y1. (Rapid to hole position #1)

N70 G43 H01 Z1. (Tool height offset #1, Rapid to initial level)

N80 G98 G81 Z-0.275 R0.1 F3. (Return to initial R-plane, drill hole #1 .275 deep 3.0 inches per minute feed)

N90 X2. (Drill hole #2)

N100 X3. (Drill hole #3)

N110 Y2. (Drill hole #4)

N120 X2. (Drill hole #5)

N130 X1. (Drill hole #6)

N140 G80 (Cancel drill cycle)

N150 M5 (Spindle stop)

N160 G28 (Return X and Y axis to home position)

N170 M30 (End program, reset the control)

%

G70 Bolt Hole Circle Program Example

This example will use a G81 and a G70 bolt hole canned cycle. The part is shown in Figure 7-10 and the parameters for a G70 are shown in Figure 7-11.

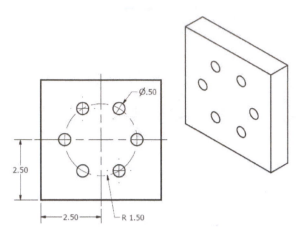

FIGURE 7–10. Part with 6 holes.

The program is shown below.

O720 (Bolt hole circle program)
N01 G90 G80 G20 G40
N02 M6 T1 (1/2 drill)
N03 G54
N04 S1200 M3
N5 G00 X2.5 Y2.5 (center position of the bolt hole circle)
N6 G43 H1 Z1.0 M8
N7 G81 G99 Z-.5 R0.1 F3.0 L0 (L0 on line 7 will cause the machine to not do this command until reading the next line, so as to not drill a hole in the center of the bolt hole circle))
N8 G70 I1.5 J60.0 L6 (G70 bolt hole circle command, I=radius of the bolt hole circle, J=starting angle for the first bolt hole from the three o' clock position, L=number of evenly spaced holes around the bolt hole circle)
N9 G90 G00 Z1.0
N10 G28
N11 M30
%

Parameters for a G70 Drill Cycle	
X*	Rapid X axis location
Y*	Rapid Y axis location
Z	Z - Depth of the hole
R	Retract Plane
F	Feed rate
L	An L0 with a G81 is used mainly for positioning. Defining an L0 is used with a G81 cycle to position to the center of the bolt hole and not drill a hole. An L with a G70 is the number of evenly spaced holes in the bolt circle.
I	Radius of the bolt circle
J	Starting position of the first hole from the three o'clock position
*Optional	

FIGURE 7–11. Parameters for a G70 canned cycle.

Peck (Deep Hole) Drilling Cycle (G83)

When deep holes are to be drilled (holes that are three to four times deeper than the diameter of the drill), a peck drilling cycle is often used. The peck drilling cycle (G83) is very similar to the G81 drilling cycle, but it uses an extra word address (Q) to specify the depth of each peck. After the drill reaches the depth of the peck, the Z axis rapids out of the hole, clearing the hole of chips (see Figure 7–12), and then pecks again and again until the drill depth is reached.

The canned peck drilling cycle will then rapid position the tool to the Z position specified by the R plane value and then back to the initial Z position. Note that the retract position is controlled by a G98 or a G99 code just like in the G81 drill cycle.

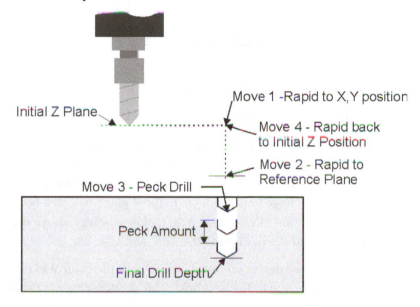

FIGURE 7–12. Canned peck drilling cycle.

Figure 7-13 shows the parameters for a G83 peck drill cycle.

G83 Peck Drill Canned Cycle	
X*	Rapid X-axis location
Y*	Rapid Y-axis location
Z	Z-depth
Q*	Pecking equal incremental depth amount (if I , J and K are not used)
I*	Size of the first peck depth if Q is not used.
J*	Amount reducing each peck after the first peck depth is Q is not used.
K*	Minimum peck depth if Q is not used.
P	Dwell time at Z-depth
R	R-plane (rapid point to start feeding)
F	Feed rate in inches per minute
* Indicates that it is Optional	

Figure 7-13. Parameters for a G83 peck drilling cycle.

The G83 canned peck drilling cycle is modal. The G83 cycle will stay active until it is canceled by a G80. If we were drilling a series of deep holes, we would only need to specify the coordinates for the next hole, just like we did with the standard G81 cycle. Figure 7–14 is a drilling example using a G83 canned peck drilling cycle.

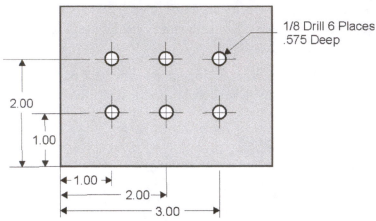

1/8 Drill 6 Places
.575 Deep

FIGURE 7–14. Canned peck drilling cycle example.

It is important to rapid the Z axis to an appropriate position above the workpiece before any canned cycle is called to establish the Z initial plane. An inappropriate Z initial plane could waste time or cause the tool to run into clamps or other obstacles. For example, if the spindle is 8 inches above the work when a drill canned cycle is called, the spindle will have to rapid down to the R plane before it begins to feed to drill the hole. The correct method is to position the Z axis to a safe, close distance above the work before the canned cycle is called so that the Initial Z will close to the work and efficient.

A G99 can be used when there are no obstacles involved. A G99 tells the control to retract the Z axis to the R plane after drilling. Then it will rapid to the next hole. This shorter retract is faster. To use this code, simply put it in the line before or in the line where the canned cycle is called, as follows:

```
O713 (canned Peck drilling)
N10 G20 (Inch mode)
N20 G00 G17 G40 G49 G80 G90
N30 T2 M6 (Tool Change to tool 2, 1/8 drill) N40 G54 (Workpiece zero setting)
N50 S1800 M3 (1800 RPM, Spindle start clockwise)
N60 G00 X1. Y1. (Rapid to hole position #1)
N70 G43 H2 Z1.0 (Tool height offset #1, Rapid to initial level)
N80 G98 G83 Z-.575 R.1 Q.3 F3.0
N90 X2.0 (Drill hole #2)
N100 X3.0 (Drill hole #3)
N110 Y2.0 (Drill hole #4)
N120 X2.0 (Drill hole #5)
N130 X1.0 (Drill hole #6)
N140 G80 (Cancel drill cycle)
N150 M5 (Spindle stop)
N160 G28 (Return X and Y axis to home position)
N170 M30 (End program, reset the control)
%
```

G73 High-Speed Peck Drill Cycle

Many machines also have a G73 high speed peck drill cycle that can be used. A G73 is faster because the spindle does not retract fully out of the hole with every peck. A G73 high speed peck drill cycle can be programmed to retract a programmed amount to clear chips or dwell to break the chips at each peck depth. Figure 7-15 shows the parameters for a G73 high-speed peck cycle.

G73 High-Speed Peck Drill Cycle	
X*	Rapid X-axis location
Y*	Rapid Y-axis location
Z	Z-depth
Q*	Pecking equal incremental depth amount (if I , J and K are not used)
I*	Size of first peck depth (if Q is not used)
J*	Amount reducing each peck after first peck depth (if Q is not used)
K*	Minimum peck depth (if Q is not used)
P	Dwell time at Z-depth
R	R-plane (rapid point to start feeding)
F	Feed rate in inches per minute
* Indicates that it is Optional	

FIGURE 7–15. Parameters for a G73 canned cycle.

Canned Tapping Cycle (G84)

When tapped holes are needed, a G84 tapping cycle can be used. The tapping cycle is commanded much like the previous canned cycles. The difference occurs when the tap reaches the programmed depth. The spindle stops and reverses itself and automatically feeds the tap out of the hole.

The feed rate of the tapping canned cycle must be coordinated with the spindle. To do this we must multiply the lead of the tap (lead equals 1 inch divided by the number of threads per inch) times the spindle RPM. The easier method is to divide the RPM by the threads per inch (TPI). For example 500 RPM divided by 20 TPI would equal 25 IPM feed rate.

When the G84 tapping cycle is commanded, the tap rapid positions to the specified X and Y coordinates and to the Z initial position. The tap then feeds down to the specified depth, cutting the threads. At the programmed depth, the spindle automatically reverses, and the tap is fed back to the R plane. On some older CNC machines, that are not capable of synchronous tapping, a floating tap holder is used to reduce the possibility of tap breakage.

The G84 canned tapping cycle is modal and will stay active until it is canceled by a G80. If we were tapping a series of holes, we would need only to specify the coordinates for the next tapped hole, as with the standard drilling canned cycles. Figure 7-16 shows the parameters for a G84 tapping cycle.

G84 Tapping Canned Cycle	
X*	Rapid X-axis location
Y*	Rapid Y-axis location
Z	Z-depth
J*	Tapping retract speed
R	R-plane (rapid point to start feeding)
F	Feed rate in inches per minute
* Optional	

Figure 7-16. Parameters for a G84 tapping cycle.

Figure 7–17 and the program that follows are an example of tapping using a G84 canned tapping cycle.

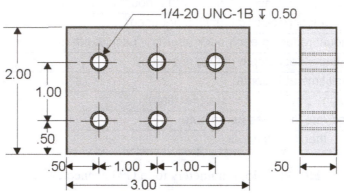

FIGURE 7–17. Sample part with six tapped holes.

O716 (Tapping)
N10 G20 (Inch mode)
N20 G00 G17 G40 G49 G80 G90
N30 T1 M6 (Spot drill)
N40 G54 (Workpiece zero setting)
N50 S3500 M3 (3500 RPM, Spindle start clockwise)
N60 G00 X.5 Y.5
N70 G43 H1 Z1. (Tool height offset 1, Rapid to initial level)
N80 G98 G81 Z-.09 R.1 F7.0
N90 X1.5 (Spot drill hole 2)
N100 X2.5 (Spot drill hole 3)
N110 Y1.5 (Spot drill hole 4)
N120 X1.5 (Spot drill hole 5)
N130 X.5 (Spot drill hole 6)
N140 G80 (Cancel canned cycle)
N150 M5 (Stop spindle)
N160 G28 (Return to reference position)
N170 M01
N190 T2 M6 (Tap drill)
N200 S4500 M3 (4500 RPM, Spindle start clockwise)
N210 G00 G90 X.5 Y.5 A0.
N220 G43 H2 Z1. (Tool height offset 2, Rapid to initial level)

N230 G98 G81 Z-.6 R.1 F8.50 (Return to initial R-plane, Drill hole 1 - .6 deep 38.5 inches per minute feed)
N240 X1.5 (Drill hole 2)
N250 X2.5 (Drill hole 3)
N260 Y1.5 (Drill hole 4)
N270 X1.5 (Drill hole 5)
N280 X.5 (Drill hole 6)
N290 G80 (Cancel canned cycle) N300 M5 (Stop spindle)
N320 G28 (Return to reference position)
N330 M01 (Optional stop)
N340 T3 M6 (TAP 1/4 -20)
N350 G00 G90 X.5 Y.5 S500 M3
N360 G43 H3 Z1.
N370 G98 G84 Z-.5 R.1 F25.0
N380 X1.5 (Tap hole 2)
N390 X2.5 (Tap hole 3)
N400 Y1.5 (Tap hole 4)
N410 X1.5 (Tap hole 5)
N420 X.5 (Tap hole 6)
N430 G80
N460 G28 (Return to reference position)
N470 M30 (End program, reset the control)

Bore Canned Cycle (G85)

A G85 canned cycle is used to bore holes. A G85 cycle will command the machine to make a rapid move from the Z initial position to the R plane. It will then bore the hole to a Z depth at the feed rate specified. It will then reverse and feed the spindle back to either the R plane or the Z initial position. Figure 7-18 shows the parameters for a G85 boring cycle.

G85 Bore in/Bore out Canned Cycle	
X*	Rapid X-axis location
Y*	Rapid Y-axis location
Z	Z-depth
R	R-plane (rapid point to start feeding)
F	Feed rate in inches per minute
* Optional	

Figure 7-18. Parameters for a G85 bore cycle.

Figure 7-19 shows an illustration of a G85 boring cycle. In the first step the spindle rapids to the hole position specified by the X and Y parameters. The spindle then rapids down to the R plane (specified by the R parameter). In step 2 the spindle feeds down to the depth specified by the Z parameter at the feed rate specified by F. In step 3 the spindle feeds back up to the R plane at the feed rate specified by the F parameter.

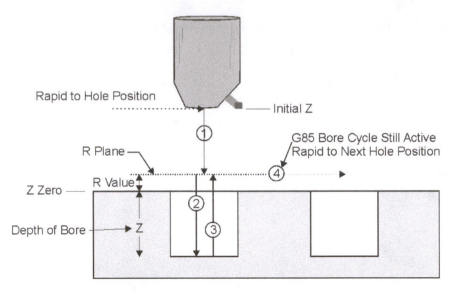

FIGURE 7–19. Example of a bore canned cycle.

The general format is

    N100 G85 Xn Yn Zn Rn Fn

G85 specifies a bore cycle.

Xn Yn are the X and Y coordinates of the hole to be bored. The X and Y can be in absolute mode (G90) or in incremental mode (G91).

Z represents the depth of the hole in absolute mode (G90) or the distance below the R plane to the hole bottom in incremental mode (G91).

R specifies the distance to the R plane in absolute mode (G90) or the distance below the initial tool position to the R plane in incremental mode (G91). If you do not specify an R plane value, the last active R plane value will be used. If there is not an R plane active, the tool will return to the initial Z position.

F specifies the feed rate.

Note that canned cycles are modal so all that would be needed to bore another hole would be the new X Y position if everything else remained the same.

Counterbore or Spotdrill Canned Cycle (G82)

A G82 canned cycle can be used to make counterbored or spotdrill holes (see Figure 7–20). A G82 cycle will command the machine to make a rapid move from the Z initial position to the R plane. It will then bore the hole to a Z depth at the feed rate specified. It will dwell for the number of seconds specified in the P value. It will then rapid the spindle back to either the R plane or the Z initial position. The parameters for a G82 are shown in Figure 7-21.

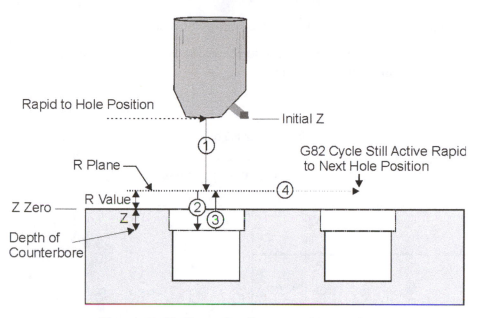

FIGURE 7–20. Example of a counterbore cycle.

G82 Spot Drill Counterbore Canned Cycle	
X*	Rapid X-axis location
Y*	Rapid Y-axis location
Z	Z-depth
P	Dwell time at z depth
R	R-plane (rapid point to start feeding)
F	Feed rate in inches per minute
* Optional	

Figure 7-21. Parameters for a G82 spot drill/counterebore cycle.

The general format of a G82 is

N100 G82 Xn Yn Zn Rn Fn Pn

G82 specifies a counterbore cycle.

Xn Yn are the X and Y coordinates of the hole to be counterbored. The X and Y can be in absolute mode (G90) or in incremental mode (G91).

Z represents the depth of the hole in absolute mode (G90) or the distance below the R plane to the hole bottom in incremental mode (G91).

R specifies the distance to the R plane in absolute mode (G90) or the distance below the initial tool position to the R plane in incremental mode (G91). If you do not specify an R plane value, the last active R plane value will be used. If there is not an R plane active, the tool will return to the initial Z position.

F specifies the feed rate.

P is the dwell time in seconds (0.01 to 99.99) at the bottom of the hole.

Note that canned cycles are modal so all that would be needed to counterbore another hole would be the new X Y position if everything else remained the same.

Helical Interpolation

Helical milling can be used to produce large internal or external threads (see Figure 7–21) or helical pockets. Helical milling involves circular interpolation in two axes (usually X and Y) plus a linear feed in the third axis (usually Z). This is especially useful for large threads.

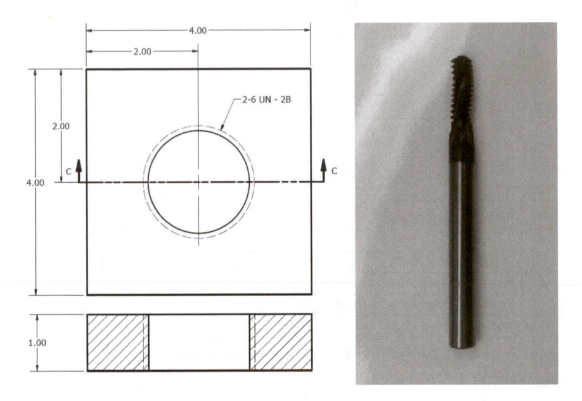

FIGURE 7–21. Examples of an internal thread that could be produced by helical milling. The right is thread milling hob.

A special type of milling cutter, called a *thread hob,* is used to mill threads. It looks like an end mill with teeth shaped like the desired thread (figure 7-21). The hob is sent to a start position. It is then fed into the workpiece and does a helical interpolation around the part. With each interpolation, it feeds the Z axis an amount equal to the lead of the thread. Lead, which is how far a thread advances in one revolution, can be calculated by dividing 1 inch by the number of threads per inch. For example, if we had a 2-inch diameter thread with 6 threads per inch, the lead would be 1 inch/6 or .166 inch. See the program example below.

Helical interpolation can be done in the X/Y plane (G17), X/Z plane (G18), or Y/Z plane (G19). The format is as follows:

X/Y plane—G17 G02/G03 X Y I J Z F

X/Z plane—G18 G02/G03 X Y I K Z F

Y/Z plane—G19 G02/G03 X Y J K Z F

G17, G18, and G19 select the plane to be used. G02 or G03 selects the direction of the helical interpolation (clockwise or counterclockwise). X, Y, and Z are the end point coordinates.

O722 (Thread Milling)

N10 G20 (Inch mode)

N20 G00 G17 G40 G49 G80 G90

N30 T1 M6 (6 Pitch x 1 inch dia. 4 flute thread hob)

N40 G54 (Workpiece zero setting)

N50 S1000 M3 (1000 RPM, Spindle start clockwise)

N60 G00 X2.0 Y-2.0

N70 G00 Z1.0 (Rapid to initial level)

N80 G01 Z-1.25 F20.0

N90 G03 X3.0 Y-2.0 R.50 F10.0 (Counter-Clockwise arc into thread)

N100 G03 X3.0 Y-2.0 I-1.0 J0.0 Z-1.083 F6.0 (Arc interpolation and hob move up one pitch on Z axis)

N110 G03 X2.0 Y-2.0 R.50 (Counter-Clockwise arc off the thread)

N120 G01 Z1.0 F20.0

N150 M5 (Stop spindle)

N160 G28 (Return to reference position)

N170 M30

%

Subprograms

Subprograms can be used to reduce redundant programming and shorten programs. For example, if you are machining a part with several holes that require multiple operations at each hole location, you would normally have to list those hole positions multiple times in the same program.

With a subprogram, you could use the subprogram to hold the hole positions and just call them for each operation. Figure 7–23 shows a part that requires the same hole pattern at four different locations. If we use incremental positioning, the holes in each pattern are the same.

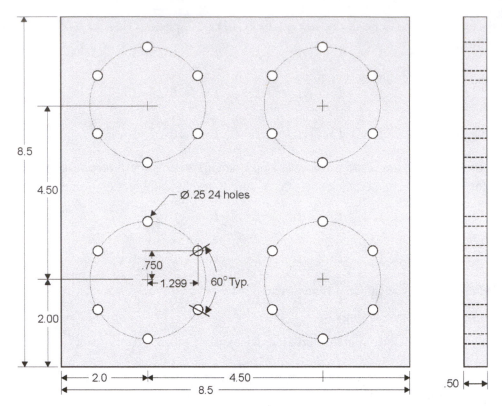

FIGURE 7–23. Part with same hole pattern in four places.

The main program is used to prepare the control. Load the tool, and get ready for machining. In line number N106 we move the tool to the hole position that is located at the 2 o'clock position on the lower left hole pattern. The first hole is drilled in line number N110. Line number N112 is used to call the subprogram. M98 is the M code that is used to call a subprogram. The subprogram's name is 1001. At this point, the control would go to program 1001 and drill the rest of the locations of the first bolt circle. Note that the locations in the subprogram are incremental. In line N70 of the subprogram, the M99 returns the control to the main program line that follows the subprogram call (line number 128). The program then continues by drilling the first hole in the next bolt pattern, calling the subroutine to finish the other holes. The process is then repeated two more times to complete all four hole patterns.

O00000 (Main Program)

N100 G20

N102 G0 G17 G40 G49 G80 G90 (Rapid, XY Plane, Cancel diameter offsets, Cancel tool length compensation, Cancel canned cycles, Absolute mode)

N104 T2 M6 (Tool change to tool 2)

N106 G54 (Workpiece coordinates are found in G54 register)

N108 S2000 M3 (Turn on spindle clockwise at a speed of 2000 RPM)

N106 G0 X3.299 Y2.75

N108 G43 H2 Z.1

N110 G99 G81 Z-.58 R.1 F4.00

N112 M98 P1001

N128 G80

N130 G90 Y7.25

N132 G99 G81 Z-.58 R.1 F4.00

N134 M98 P1001

N150 G80

N152 G90 X7.799Y7.25

N154 G99 G81 Z-.58 R.1 F4.00

N156 M98 P1001

N172 G80

N174 G90 Y2.75

N176 G99 G81 Z-.58 R.1 F4.00

N178 M98 P1001

N194 G80

N196 M5

N198 G91 G28 Z0.

N200 G28 X0. Y0.0

N202 M30 (Program end, Memory reset)

O1001 (Subprogram)

N10 G91 (Incremental mode)

N20 X-1.299Y.75

N30 X-1.299Y-.75

N40 Y-1.5

N50 X1.299Y-.75

N60 X1.299Y.75

N70 M99 (Return from subroutine to the main program)

%

Chapter Questions

1. What type of information does the programmer get from the part drawing?

2. What must be taken into consideration when deciding on the machine to be used?

3. What is a process plan?

4. Describe the setup sheet.

5. What are the advantages of using a canned cycle?

6. List 2 of the more commonly used canned cycles.

7. Calculate the feed rate for a 1/2-13 tap running at 600 RPM.

8. Using the tool data tables shown in Figure 7–24, develop a process plan (Figure 7-25) and then write a program to execute the profile milling and the hole operations for the part drawing shown in Figure 7–26. Use absolute programming and cutter compensation. Mill the part contour .50 deep.

Tool #	Operation	Tool	Speed (RPM)	Feed (IPM)
1	Mill Profile	.750 End Mill	150	6.00
2	Tap Drill	5/16 Drill	1000	3.00
3	¼ Drill	¼ Drill	1200	2.5
4	Tap	3/8-16 Tap	500	31.25

FIGURE 7–24. Tools available to program the part shown in Figure 7–16.

Acme Machining Inc.- Process Plan				Part #
Operation	Tool #	Tool Description	RPM	Feed rate

FIGURE 7–25. Process planning sheet for question 8.

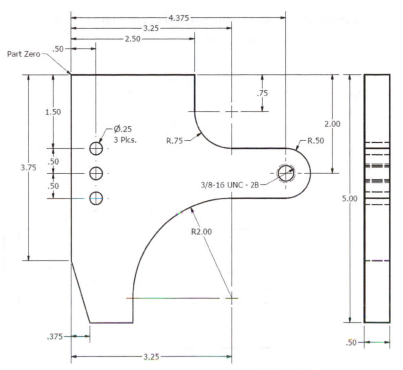

FIGURE 7–26. Part for question 8.

9. Write a program to machine the part shown in Figure 7–27. Use canned cycles where appropriate.

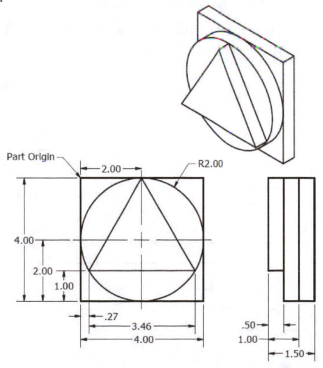

FIGURE 7–27. Part for question 9.

10. Write a program to machine the part shown in Figure 7–28. Use canned cycles where appropriate. Mill the contour .5 deep

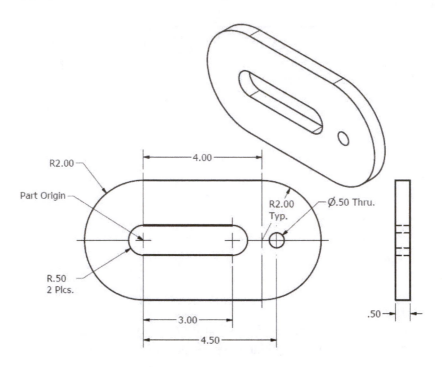

FIGURE 7–28. Part for question 10.

11. Write a program to machine the part shown in Figure 7–29. Use canned cycles where appropriate. Mill the contour .5 deep

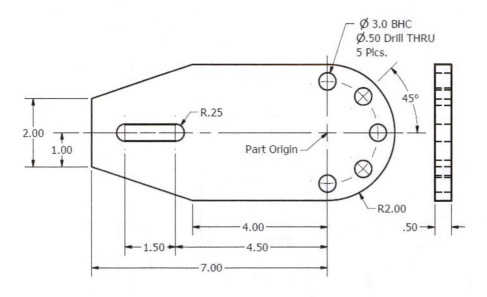

FIGURE 7–29. Part for question 11.

12. Write a program to machine the part shown in Figure 7-30. Use canned cycles where appropriate. Mill the contour .5 deep

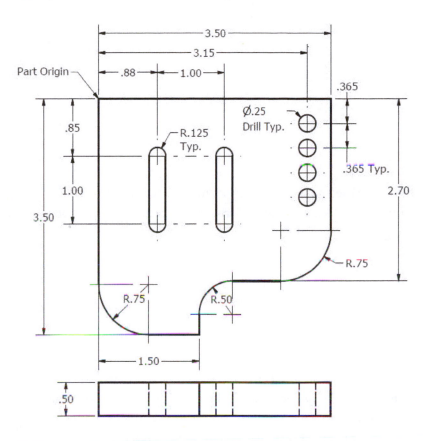

FIGURE 7–30. Part for question 12.

Chapter 8

CNC TURNING MACHINES

INTRODUCTION

A CNC lathe is commonly called a turning center. Its main function is to create high-quality cylindrical parts efficiently. Turning centers can machine internal and external surfaces. Other machine operations, such as drilling, tapping, boring, and threading, are also done on turning centers.

OBJECTIVES

Upon completion of this chapter, the reader will be able to:
- Name the major components of the turning center.
- Correctly identify the major axes on turning centers.
- Describe the three major work-holding devices used on CNC lathes.
- Describe the common machining operations and the tools associated with them.
- Explain tool-wear offsets.
- Describe how geometry offsets or workpiece coordinates are set.
- Correctly identify safe working habits associated with CNC lathes.
- Identify and explain turning center controls.

Introduction to Turning Centers

The turning center is able to produce cylindrically shaped parts in great volumes and with incredible accuracy. The first numerically controlled turning machines were developed in the mid-1960s and were little more than an engine lathe retrofitted with a control and drive motors. Today's turning centers can be equipped with dual tool turrets, dual spindles, milling head attachments, and a variety of other specialized features to make them capable of machining even the most complex parts in one setup.

Types of CNC Turning Machines

The standard flat-bed configuration is still evident on some CNC lathes; however, most turning machines today have a slant-bed configuration (see Figure 8–1). Slanting the bed on a CNC lathe allows the chips to fall away from the slideways and allows the operator easy access to load and unload parts.

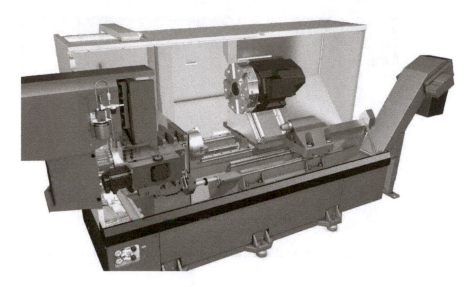

FIGURE 8–1. Slant-bed-style turning center. Courtesy Haas Automation Inc.

Components of CNC Lathes

The main components of a CNC lathe or turning center are the headstock, tailstock, turret, bed, and carriage (see Figure 8–2).

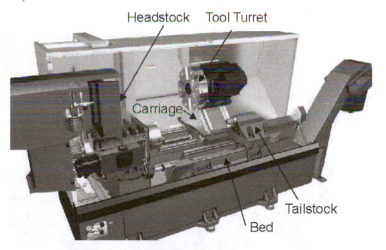

FIGURE 8–2. Slant-bed lathe components. Courtesy Haas Automation Inc.

Headstock

The headstock contains the spindle and transmission gearing, which rotates the workpiece. The headstock spindle is driven by a variable speed motor. The spindle motor delivers the required horsepower and torque through a drive belt or series of drive belts. Figure 8–3 shows a cutaway view of a typical spindle.

FIGURE 8–3.   Cutaway view of a headstock and spindle. Courtesy Haas Automation Inc.

Tailstock

The tailstock is used to support one end of the workpiece. The tailstock slides along its own set of slideways on some turning machines and on the same set of slideways as the carriage on conventional-style CNC lathes. The tailstock has a sliding spindle much like that of the tailstock on a manual lathe. Two types of tailstocks are available: manual and programmable. The manual tailstock is moved into position by the use of a switch or hand wheel. The programmable tailstock can be moved manually or can be programmed like the tool turret (see Figure 8–4).

FIGURE 8–4.   Programmable tailstock. Courtesy Haas Automation Inc.

Tool Turrets

Tool turrets on turning machines come in all styles and sizes. The basic function of the turret is to hold and quickly index cutting tools. Each tool position is numbered for identification (see Figure 8–5). When the tool needs to be changed, the turret is moved to a clearance position and indexes, bringing the new tool into the cutting position. Most tool turrets can move bi-directionally to assure the fastest tool indexing time. Tool turrets can also be indexed manually, using a button or switch located on the control panel.

FIGURE 8–5. Twelve-station tool turret. Each turret position is numbered for identification purposes. Courtesy Haas Automation Inc.

Bed

The bed of the turning center supports and aligns the axis and cutting tool components of the machine. The bed is made of high-quality cast iron and will absorb the shock and vibration associated with metal cutting. The bed of the turning center lies either flat or at a slant to accommodate chip removal. The slant of the bed is usually 30 to 45 degrees (see Figure 8–6).

FIGURE 8–6. The slant bed is designed for quick chip removal and easy operator access. Courtesy Haas Automation Inc.

Carriage

The carriage slides along the bed and controls the movement of the tool. The basic CNC lathe has two major axes, the X and Z axes (see Figure 8–7).

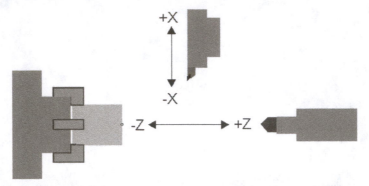

FIGURE 8–7. Lathe axes of motion. Note that the Z axis is always in line with the spindle.

The Z axis is in the same plane as the spindle, just as on machining centers. The X or cross-slide axis runs perpendicular to the Z axis. Negative Z axis (–Z) motion moves the tool turret closer to the headstock. Positive Z (+Z) motion moves the turret or tool away from the headstock or toward the tailstock. Negative X (–X) motion moves the tool or turret toward the centerline of the spindle, and positive X (+X) motion moves the tool or turret away from the centerline. Some CNC turning centers are equipped with a programmable tailstock or third axis known as the W axis. More complex turning centers have four axes and have two opposing turrets. The turrets are independent of one another and can perform different machining operations at the same time (see Figure 8–8).

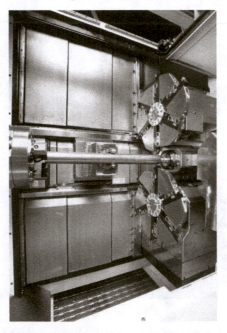

FIGURE 8–8. Two independent turrets allow two machining operations to be done simultaneously. Courtesy Haas Automation Inc.

Work Holding

Work-holding devices are an integral part of a CNC lathe. As the demand for higher production increases, so does the need for an understanding of the different types of work-holding devices. The most common work-holding method used on turning machines is the chuck.

Chucks

There are many different types of chucks, but the most common type is the three-jaw, self-centering, hydraulic chuck (see Figure 8–9).

FIGURE 8–9. Three-jaw hydraulic chuck.

This type of chuck has three jaws, all of which move in unison under hydraulic power. The chucks are activated by a foot switch that opens and closes the chuck jaws.

There are two types of chuck jaws: hardened jaws and soft jaws. Hardened jaws are used where maximum holding power is needed on unfinished surfaces.

Soft jaws are used on parts that cannot have much runout or on finished surfaces that cannot be marred. Soft jaws are typically made from soft steel and are turned to fit each type of part. Soft jaws can be machined with ordinary carbide tooling.

Collet Chucks

The collet chuck is an ideal work-holding device for small parts where high accuracy is required. The collet chuck assembly consists of a draw tube and a hollow cylinder with collet pads. Collets are available for holding hexagonal, square, and round stock. They are typically used with bar feed systems.

Fixtures

Fixtures are work-holding devices used for odd-shaped or other hard to hold workpieces. Fixtures may be held in the chuck or can be bolted directly to the spindle. Fixtures are spun in the spindle, so they must be balanced. Unbalanced fixtures can cause severe damage to the machine and possibly injure the operator.

Material Handling

Material-handling devices increase production rates and reduce labor costs. The types of devices used are determined by part size, shape, and production levels.

Bar Feeders

Bar feeders automatically load rough stock into the work-holding device. The raw stock is fed into the machine by pneumatic or hydraulic pressure. The stock is fed the same distance each time through the use of stock stops. When the stock reaches the stock stop, the work-holding device closes and clamps the workpiece in place. Bar feeders eliminate the need for the operator to manually load individual part blanks (see Figure 8–10).

FIGURE 8–10. Bar-feeding mechanisms are capable of handling full-length bars of stock, which are sometimes 20 feet in length. Courtesy Haas Automation Inc.

Part Loaders and Unloaders

Individual part blanks can be automatically loaded using part loaders. Part loaders take up less space than bar feeders, but the parts have to be cut to length prior to loading. A part loader is an auxiliary arm that places the precut stock into the chuck or collet. The auxiliary arm can also unload parts after the necessary machining has been done.

Robotic Loading Systems

The use of robotic equipment is increasing. Robots can be used to load and unload parts, retrieve parts from pallets, and change chuck jaws (see Figure 8–11).

Robots can communicate directly with the CNC controls or with switches and sensors. The robot controllers are designed to be easy to program and integrate with a CNC control.

FIGURE 8–11. Robotic handling systems increase the productivity of turning centers.

Conveyors and fixtures are sometimes needed to use robotic loading systems. Grippers need to be designed and built so that the robot can handle different part shapes. Some robots have automatic tool-changing capability so that they can even change grippers automatically as required to pick up different parts.

Parts Catchers

Parts catchers are used on small-diameter workpieces. Parts catchers consist of a tray that, prior to the tool cutting off the part, tips forward and catches the completed part and delivers the part to the outside of the machine.

Chip Conveyers

Chip conveyers automatically remove chips from the bed of the machine. The chips produced by machining operations fall onto the conveyer track and are transported to scrap or recycling containers (see Figure 8–12).

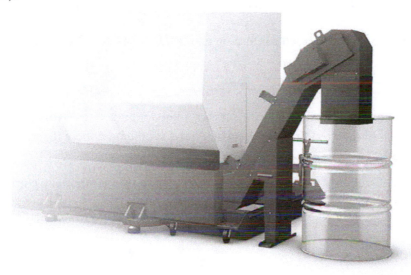

FIGURE 8–12. Chip conveyers provide automated chip disposal to provide a chip-free work environment.

Cutting Tools

Turning centers use tool holders with indexable inserts. The tool holders on CNC machines come in a variety of styles, each suited for a particular type of cutting operation. The machining operations discussed in this chapter include facing, turning, grooving, parting, boring, and threading. Chapter 2 covered tool holders and carbide inserts.

Facing

Facing operations involve squaring the face or end of the stock. The tool needs to be fed into the stock in a direction that will push the insert toward the pocket of the holder (see Figure 8–13).

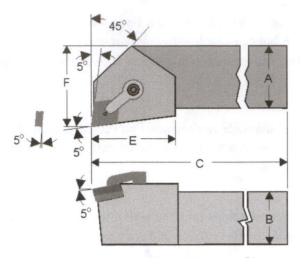

Style L Tool Holder with an 80° Insert

FIGURE 8–13. Tool style L can be used for turning and facing using an 80-degree diamond insert.

Turning

Turning operations remove material from the outside diameter of the rotating stock. Rough turning removes the maximum amount of material from the workpiece and should be done with an insert with a large included angle.

The large included angle will insure that the tool has the proper strength to withstand the cutting forces being exerted. Profile turning uses an insert with a smaller included angle. If the finish profile requires the use of a small, sharp-angled insert, a series of semi-finish passes are necessary to insure against tool breakage (see Figure 8–14).

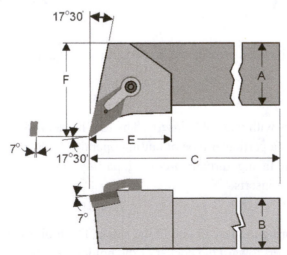

Style Q Tool Holder with an 55° Insert

FIGURE 8–14. Tool style Q is used for profile turning using a 55-degree diamond insert.

Grooving

For internal and external grooving, the tool is fed straight into the workpiece at a right angle to its centerline. The cutting insert is located at the end of the tool (see Figure 8–15). Grooving operations include thread relieving, shoulder relieving, snap-ring grooving, O-ring grooving, and oil reservoir grooving.

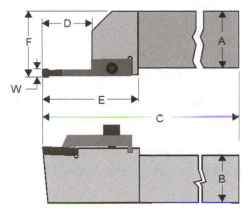

Style NG Grooving Tool Holder

FIGURE 8–15. Tool style NG for grooving.

Parting

Parting is a machine operation that cuts the finished part off from the rough stock. This operation is similar to grooving. The tool is fed into the part at a right angle to the centerline of the workpiece and is fed down past the centerline of the work, thus separating it from the rough stock. The parting tool has a carbide insert located at the end of the tool and has a slight back taper along the insert for clearance (see Figure 8–16).

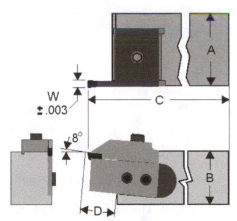

Style KGSP Parting Tool Holder

FIGURE 8–16. Tool style KGSP for parting.

Boring

Boring is an internal turning operation that enlarges, trues, and contours previously drilled or existing holes. Boring is done with a boring bar (see Figure 8–17).

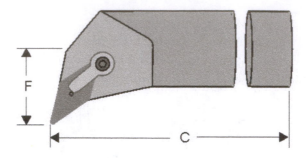

Style LP Boring Bar with 55° Diamond Insert

FIGURE 8–17. LP-style boring bar using a standard 55-degree diamond insert.

Threading

Threading is the process of forming a helical groove on the outside or inside surface of a cylindrical part. Threads can be cut in multiple ways, but for this tooling section we will concentrate on single-point threading tools (see Figure 8–18).

Single-point threading tools are typically 60-degree carbide inserts clamped in a tool holder. The threading tool is fed into the work and along the part at a feed rate equal to the lead of the thread. (The lead of a thread is the distance it advances in one revolution.) The lead can be calculated by dividing 1 inch by the number of threads. For example, if you had eight threads per inch, the calculation would be 1/8 or .125.

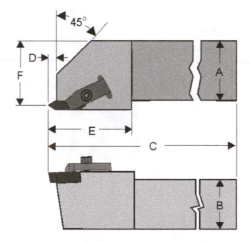

Style NS Threading Tool

FIGURE 8–18. NS-style threading tool. This style holder can be equipped with different angled threading inserts for different types of threads.

Presetting Tools

Presetting tooling involves setting the cutting point of the tool in relation to a predetermined dimension. Preset tools enable the operator to just replace a dull tool with a new tool that is already set to the correct dimensions so cutting can resume without changes. On some machines, tool dimension setting is done with a toolset arm (see Figure 8–19).

The operator moves each tool close to the sensor and touches off the tool on the X and Z sensor to set the dimension. When the tool tip comes in contact with the sensors, the offset dimension is recorded in an offset page. When all the tools are measured, the operator uses one premeasured tool and touches off on the end of the workpiece. The control then uses this information to determine where the workpiece is located and also calculates and adjusts for the different tool tip locations. Some controls can use the tool measurement arm to check tools automatically between cutting operations to be sure tools have not been worn, damaged, or broken during cutting.

FIGURE 8–19. The tool presetting arm greatly reduces the time required to measure tools.

Machines that are not equipped with tool presetting arms use standard geometry offsets, discussed later in this chapter.

Machine Reference Position, Tool Change Position, and Part Origin

CNC turning centers have three zero points. Figure 8–20 shows the typical zero points for a CNC turning center.

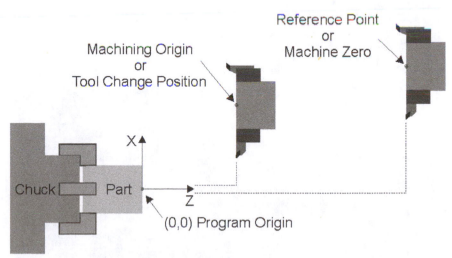

FIGURE 8-20. Zero points for a CNC turning center.

Machine Zero

The machine zero point is set by the machine manufacturer. It is the point at which all of the axes are zeroed out.

Tool Change Position

The tool change position is also called the machining origin. The tool change position is a safe location that the machine returns to when indexing to a new tool.

The tool change position is determined during program setup. The location is input at the beginning of a program by use of the zero offset command (G50). During operation the machine will execute all movements relative to this origin. It should be noted that on small turning centers the machine zero is often used as the tool change position. On larger machines it may take too much time to return to the machine zero to change a tool so a tool change position that is closer to the chuck is used.

Part Origin

The part origin can also be called a program origin. The part origin is the zero point from which all of the part program dimensions were created. When a CNC turning center is setup, the operator uses tool offsets to locate the program origin from the machining origin.

Offsets

There are two kinds of offsets used on CNC turning machines: tool offsets and geometry offsets.

Tool Offsets

Tool offsets, also called tool-wear offsets, are an electronic feature for adjusting the length and diameter of machined surfaces. CNC turning machines have offset tables in which the operator can input or change numbers to adjust part sizes without changing the program. Adjusting the offsets is a responsibility of the CNC machine operator. If the part sizes don't meet the part print requirements, the operator alters the offset, for the tool in the tool-wear offset table (see Figure 8–21).

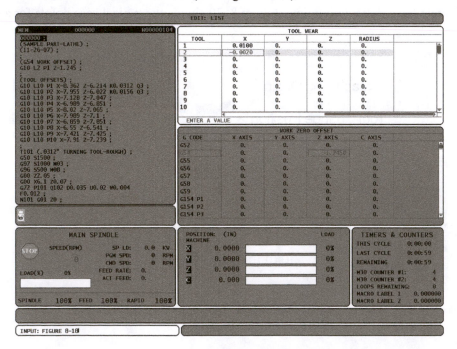

FIGURE 8-21. Tool offset table. Notice that this is a wear offset table.

Geometry Offsets

Geometry offsets, or workpiece coordinates, are used to tell the control where the workpiece is located. The workpiece coordinate is the distance from the tool tip, at the home position, to the workpiece zero point. The workpiece zero point is normally located at the end and center of the workpiece or at the chuck face and center of the machine (see Figure 8–22).

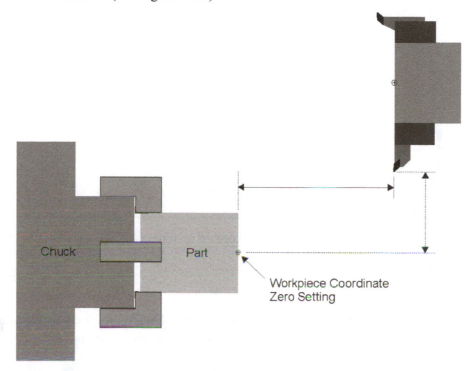

FIGURE 8–22. Tool geometry offsets, also known as workpiece coordinate settings, are generally set as the distance from machine home to the end and center of the workpiece.

Geometry offsets, or workpiece coordinates, can be registered two different ways. The most common approach is to use a preparatory or G-code such as a G50 or a G92. The offset distance is determined by touching the tool tip off on the workpiece and recording the distance.

The position or distance can be determined accurately by using the position screen on the control. For example, if the distance from the home position to the face of the workpiece is 16.500 inches and the distance from the tool tip to the centerline of the workpiece is 8.500 inches, the G50 or G92 would be G92 X8.500 Z16.500.

To find the centerline of the workpiece, the operator takes a skim cut off of the outside of the workpiece. The operator then measures the turned diameter and adds that dimension to the X-axis machine position.

If a geometry offset is used, the X and Z values would be loaded into the geometry offset table under the tool offset number, and the G50 or G92 would not be needed. It is important to remember that every tool that is used in the program will need to be measured in this manner. Figure 8–23 shows a tool offset table.

No.	X Axis	Z Axis	Radius	Tip	Machine Position (Relative)
					Tool Offset (Geometry) N0000
01	00.000	00.000	00	00	
02	00.000	00.000	00	00	X00.0000
03	00.000	00.000	00	00	Y00.0000
04	00.000	00.000	00	00	
05	00.000	00.000	00	00	
06	00.000	00.000	00	00	
07	-12.346	-8.567	.032	3	
08	-6.5671	-6.987	.015	2	(Inch)
09	-4.5672	-3.7865	.032	3	
10	-3.789	-4.8923	.032	3	
11	00.000	00.000	00	00	
12	-10.500	-4.876	.031	1	
13	-9.5624	-5.8763	.015	3	
14	00.000	00.000	00	00	
15	00.000	00.000	00	00	
16	-8.5390	-7.9845	.015	3	

FIGURE 8–23. Tool offset table.

Machine Control Operation

Safety

Before you operate any machine, remember that no one has ever thought they were going to be injured. But it happens! It can happen in a split second when you are least expecting it, and an injury can affect you for the rest of your life.

You must be safety minded at all times. Please get to know your machine before operating any part of the machine control and please keep these safety precautions in mind.

- Wear safety glasses and side shields at all times.
- Do not wear rings or jewelry that could get caught in a machine.
- Do not wear long sleeves, ties, loose fitting clothes, or gloves when operating a machine. These can easily get caught in a moving spindle or chuck and cause severe injuries.
- Keep long hair covered or tied back while operating a machine. Many severe accidents have occurred when long hair became entangled in moving tooling and machinery.
- Keep hands away from moving machine parts.
- Use caution when changing tools. Many cuts occur when a wrench slips.
- Stop the spindle completely before doing any setup or piece loading and unloading.
- Do not operate a machine unless all safety guards are in place.
- Metal cutting produces very hot, rapidly moving chips that are very dangerous. Long chips are especially dangerous. You should be protected from chips by guards or shields. You must always wear safety glasses with side shields to prevent chips from flying into your eyes. Shorts should not be worn because hot chips can easily burn your legs. Hot chips that land on the floor can easily burn though thin-soled shoes.

- Many injuries occur during chip handling. Never remove chips from a moving tool. Never handle chips with your hands. Do not use air to remove chips. They are dangerous when blown around and can also be blown into areas of the machine where they can damage the machine.
- Securely clamp all parts. Make sure your setup is adequate for the job.
- Use proper methods to lift heavy materials. A back injury can ruin your career. It does not take an extremely heavy load to ruin your back; bad lifting methods are enough.
- Safety shoes with steel toes and oil-resistant soles should be worn to protect your feet from dropped objects.
- Watch out for burrs on machined parts. They are very sharp.
- Keep tools off of the machine and its moving parts.
- Keep your area clean. Sweep up chips and clean up any oil or coolant that people could slip on.
- Use proper speeds and feeds. Reduce feed and speed if you notice unusual vibration or noise.
- Dull or damaged tools break easily and unexpectedly. Use sharp tools and keep tool overhang short.

Machine Controls

As you look at a turning center control, you see what seems to be an endless number of buttons, keys, and switches (see Figure 8–24). Although every manufacturer has its own style of control, they all have basically the same features. If you have a good understanding of one machine, the next control will be much easier to learn.

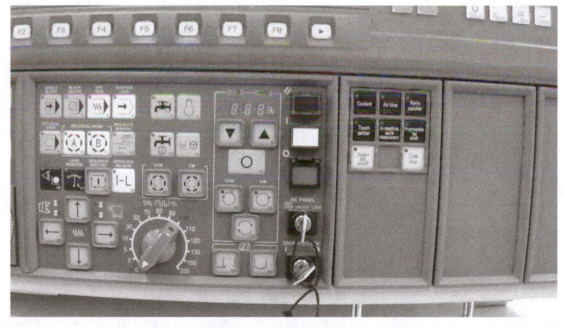

FIGURE 8–24. Typical CNC Lathe control configuration.

Manual Control

Manual control features are buttons or switches that control machine movement (see Figure 8–25).

FIGURE 8–25. Manual machine control features.

Emergency Stop Button

The emergency stop button is the most important component of the machine control. This button has saved more than one operator from disaster. The emergency stop button, which shuts down all machine movement, is a large red button. Emergency stop buttons should be used when it is evident that a collision or tool breakage is going to occur. Emergency stop buttons are often located in more than one area on the machine tool and should be located prior to operating a machine.

Moving the Axes of the Machine

Manual movement of the machine axes is done a number of different ways. Most controls are equipped with a pulse-generating hand wheel (see Figure 8–26).

FIGURE 8–26. This hand wheel could be used to move an axis.

The hand wheel gives the operator a great deal of control of the machine axes.

The hand wheel has an axis selection switch that allows the operator to choose which axis he or she wants to move. The handle sends a signal or electronic pulse to the motors, which move the carriage and the cross slide.

Machines are also equipped with jog buttons (see Figure 8–27). When the jog button for an axis is pressed, the axis moves. The distance or speed at which the machine moves is selected by the operator prior to the move.

FIGURE 8–27. Axes jog buttons.

Some machines have a joystick that is used to move the machine axes in the direction that the joystick is moved. For example, if the joystick is moved to the down position, the cross slide moves down; if the joystick is moved to the left, the carriage moves toward the headstock, and so on. The distance or speed at which the machine moves is selected in much the same way as the jog buttons. The selection options include rapid traverse, selected feed rate, or incremental distance.

Cycle Start/Feed Hold Buttons

The two most commonly used buttons on the control are the cycle start and feed hold buttons (see Figure 8–28). The cycle start button is used to start execution of the program. The feed hold will stop execution of the program without stopping the spindle or any other miscellaneous functions. By pushing cycle start, the operator can restart the execution of the program.

FIGURE 8–28. Cycle start and feed hold buttons.

Spindle Speed and Feed Rate Override Switches

Spindle speed and feed rate overrides are used to speed up or slow down the feed and speed of the machine during cutting operations (see Figure 8–29). The override controls are typically used by the operator to adjust to changes in cutting conditions, such as hard spots in the material. Feed rates can generally be adjusted from 0 to 150 percent of the programmed feed rate. Spindle speeds can often be adjusted from 0 to 200 percent of the programmed spindle speed.

FIGURE 8–29. Feed rate and spindle speed override controls.

Single Block Operation

The single block option on the control is used to advance through the program one block at a time. When the single block switch is on, the operator presses the button each time he or she wants to execute a program block. When the operator wants the program to run automatically, he or she can turn the single block off and press cycle start, and the program will run through without stopping. The single block switch allows the operator to watch each operation of the program carefully. It is often used to test a new program.

Manual Data Input

Manual data input or MDI is an input method that can be used for making changes to a previously loaded program or as a means of inputting data for the machine to act on manually, especially for setup purposes. MDI is done through the alphanumeric keyboard located on the control (see Figure 8–30). The keyboard is made up of letters, numbers, and symbols. These keys allow the operator to input a series of commands or a whole program. MDI is often used by the operator to input and execute a single line of code: such as M03 S800 (turn the spindle on clockwise at 800 RPM).

FIGURE 8–30. An alphanumeric keyboard, which contains a wide variety of keys, allows the operator to input information.

Program Editing

When a part program is written, it may have some errors or need some changes. These errors in programming usually show up on the shop floor. The operator or programmer can make changes at the machine control using the program edit mode. The programmer uses the display screen to locate the program errors and the keyboard to correct the errors.

Display

The display shows information such as the program or part graphics. In most cases, the program is too long to fit on the screen and is separated into pages. The page or cursor button allows one to move through consecutive parts of the program. Graphics are also displayed on the screen. Graphics are a representation of the part and the tool path generated by the active program.

Diagnostics

The diagnostics mode consists of several routines that detect errors in the machine system. If errors exist, the error number and message will be displayed on the screen. The error message will help prompt the operator or service technician to the cause of the problem.

Conversational Programming

Conversational programming enables an operator to write a part program by answering questions. The questions guide the programmer through each phase of machining operations such as part material, turning, threading, or grooving. After each response to a question, further questions are presented until the operation is complete. This type of manual data input is quicker than using word address programming. There is no standard conversational part programming language, and each manufacturer's conversational programming is quite different. Once the conversational program is complete, most controls will transform the conversational language into a standard word address program. At that point the operator would have to make edits in the word address program.

CHAPTER QUESTIONS

1. Name four of the main components that make up the CNC turning center.

2. What is the purpose of the tool turret?

3. State one advantage of a slant-bed-style CNC turning center over the flat-bed-style CNC lathe.

4. Which of the two major axes associated with the turning center always lies in the same plane as the spindle?

5. What is the most common type of work-holding device used on the turning center?

6. Describe threading.

7. Which machining operation cuts the finished part off the rough stock?

8. When are tool-wear offsets used?

9. What is a geometry offset?

10. How can an operator accurately judge the position of the tool?

11. From what location is the workpiece zero or geometry offset typically calculated?

12. What is a bar feeder?

13. How can chips be automatically removed from the bed of the turning center?

14. Name two manual control devices that allow the operator to move the axes of the machine.

15. What tasks do override switches perform?

Chapter 9

PROGRAMMING CNC TURNING Centers

INTRODUCTION

This chapter covers the steps necessary to properly plan, program and set up a part for turning. Turning centers have the capabilities to reduce part programming time and increase part quality through the use of canned cycles and tool-nose radius compensation.

OBJECTIVES

Upon completion of this chapter, the reader will be able to:

- Identify the two main axes of movement associated with a turning center.
- Describe the recommended sequence of operations for a turning center.
- Program parts using linear and circular moves.
- Define the term "tool-nose radius compensation."
- Explain the use of tool-nose direction vectors.

Turning Centers

The turning center uses two basic axes of motion and the machining center uses three. The two basic axes of movement on the turning center are the X axis, which controls the diameter of the part, and the Z axis, which controls the length.

The X axis is normally programmed with diameter values rather than radius values, so the actual position of the tool would be the radial distance from the centerline (see Figure 9–1). The spindle centerline would be an X0 position. When positioning the tool in the X axis of travel, we will seldom position to a –X position, except in the case of a facing operation. The Z axis part origin or zero position can be either at the right end of the part or a position located near the spindle of the machine.

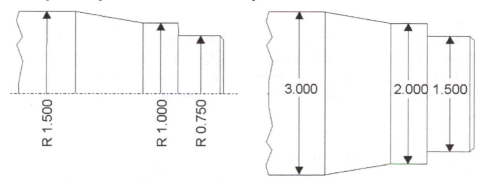

FIGURE 9–1. CNC lathes are programmed in diameter values.

Planning the Program

When planning a program for the turning center, we need to be aware of the tooling, the work-holding device, and the part print. The part drawing or part print gives the programmer detailed information about requirements. The shape of the part, part tolerances, material requirements, surface finishes, and the quantity of parts all have an impact on the program, as well as where the part zero or datum will be located.

Work Holding

Work-holding devices on turning centers will typically be chuck jaws, collets, or centers. A typical CNC turning center will have a hydraulically actuated chuck. A foot pedal, located in front of the machine, will open and close the chuck jaws or collet. Chuck jaws are either soft or hardened. Hardened chuck jaws are used for maximum holding power, but they will sometimes mar the surface of the parts. Parts that cannot be marred or that must be concentric can be held in soft jaws, made of soft, low-carbon steel or aluminum and usually machined to fit the workpiece.

Chuck jaws need to be repositioned quite often for different size workpieces. When the jaws are repositioned, you must be sure that the jaws are tight and that they are all the same distance from the centerline of the chuck. This insures the part will run true. The maximum travel of the chuck jaws is usually no more than 3/8 of an inch, which is why some machine shops use collets.

Collets are available in standard fractional sizes and can be changed quickly for various size parts. They are also used when greater part concentricity is needed. Chucked parts tend to run out, especially parts held in hardened chuck jaws. All work-holding information needs to be included in the process plan or setup sheet to insure the reliability of the program.

Tooling Considerations

The necessary tooling is determined by the configuration of the part. The most common tools used on turning centers are 80-degree diamond inserts for roughing work, 35-degree diamond inserts for finishing, grooving inserts, and 60-degree threading inserts.

The type and style of tool holders used will be based on the clearance conditions, which are dictated by the part configuration and the amount of stock to be removed. The tools and tool-holder styles needed will also have to be chosen and entered into the setup sheet or process plan.

The Process Plan

Process planning involves deciding when certain turning operations will take place. Primary machining operations are those operations that will take place on the CNC machine. Although the part configuration will have a decided effect on the sequence of operations, there are some general rules to follow when deciding on the sequence of machining operations. The recommended procedures for turning, threading, and grooving are as follows:

- Facing
- Rough turning of the profile of the part
- Drilling
- Rough boring
- Finish turning of the profile of the part

- Finish boring
- Grooving
- Threading

Roughing operations should be performed when the maximum amount of material is still in place to insure that the part will not flex or deflect away from the cutting tool. Process planning is usually done by the programmer and/or operator in smaller job-shop settings. In large shops, the process plan would come down from the engineering area and would include information for each step in machining the part. In a small job shop, the operator becomes the manufacturing engineer and determines the best, most economical way to produce the part. The operator determines the operating sequence, types of cutting tools, work-holding devices, and machining conditions (cutting speed, feed rate, and depth of cut). Whether process planning is done by the manufacturing engineer or the operator, a plan for each setup needs to be developed. The process plan is done on a process planning sheet (see Figure 9–2).

Acme Machining Inc. - Process Plan			Part Datum		Part #
Operation	Tool #	Tool Description		RPM	Feed Rate

FIGURE 9–2. The process plan outlines the machining steps to be done on the part.

The Setup Sheet

The job setup sheet is a detailed explanation of how the parts are to be set up, the type of work-holding device to be used, where the part datum is located, and the type of tools to be used (see Figure 9–3).

ACME Machining Inc. – Job Setup Sheet		Part Number ACME 157B12
Notes: Use Machine # 20. Use calibrated tools from the machine's tooling cabinet. Replace		
Part Datum:		
Tool Number	Tool Description	Operation
7	80 Degree Diamond	Rough Turn
2	35 Degree Diamond	Finish Profile
4	.125 Grooving Tool	Groove
12	60 degree Thread Tool	Thread
8	1" Drill	1" Hole

FIGURE 9–3. The CNC turning center setup sheet may include a sketch for clarity. Setup sheets help assure consistent quality and rapid setups. They help assure that whoever sets up a job and runs it does it in the same manner each time.

The setup sheet is useful to reference if the same or similar parts are programmed in the future. The setup sheet should contain all of the necessary information for the job. Setup sheets can dramatically reduce setup times. Notes added by the operator can be very valuable the next time the part is run.

Word Address Programming for Turning Centers

The word address programming format is a system of characters arranged into blocks of information. G-codes (preparatory functions), and M-codes (miscellaneous functions), are the main control codes in word address programming. Figures 9–4 and 9–5 list the more commonly used M- and G-codes found in industry. Use these or the codes for your particular turning center as a reference when completing the exercises in this chapter.

M00	Program stop
M01	Optional stop
M02	End of program
M03	Spindle start clockwise
M04	Spindle start counterclockwise
M05	Spindle stop
M07	Mist coolant on
M08	Flood coolant on
M09	Coolant off
M30	End of program & reset to the top of program
M98	Subprogram call
M99	End subprogram & return to main program

FIGURE 9–4. Commonly used turning center M (miscellaneous) functions.

G00	Rapid traverse (rapid move)
G01	Linear positioning at a feed rate
G02	Circular interpolation clockwise
G03	Circular interpolation counter-clockwise
G04	Dwell
G09	Exact stop
G10	Programmable data input
G20	Input in inches
G21	Input in mm
G22	Stored stroke check function on
G23	Stored stroke function off
G27	Reference position return check
G28	Return to reference position
G32	Thread cutting
G40	Tool diameter compensation cancel
G41	Tool diameter compensation-left
G42	Tool diameter compensation-right
G50	Max. spindle speed clamp
G70	Finish machining cycle
G71	Turning cycle
G72	Facing cycle
G73	Pattern repeating cycle
G74	Peck drilling cycle
G75	Grooving cycle
G76	Threading cycle
G80	Canned cycle cancel
G81	Drill caned cycle
G82	Spot drill canned cycle
G83	Peck drilling caned cycle
G84	Tapping canned cycle
G85	Boring canned cycle
G86	Bore and stop canned cycle
G95	Feed per revolution
G96	Constant surface speed control
G97	Constant surface speed control cancel

FIGURE 9–5. Commonly used turning center G-codes or preparatory functions.

Review of Programming Procedures

Whether you are programming a machining center or a turning center, all CNC programs basically follow a general order.

The general order is as follows:
- Start-up procedures
- Tool call
- Workpiece location block
- Spindle speed control
- Tool motion blocks
- Home return
- Program end procedures

Startup (Preliminary) Procedures

Startup procedures are those commands and functions that are necessary at the beginning of the program. A standard startup procedure usually involves cancelling tool compensation, absolute or incremental programming, standard or metric, and the setting of the work-plane axis. Every manufacturer of machine tools has its own suggested format for startup. Below is an example of typical turning center startup procedure blocks.

N0001 G90; Absolute programming.

N0002 G20 G40; Inch units, tool-nose radius compensation cancel.

Tool Change and Tool Call Block

An M06 tool change code is not used on the turning center for changing or selecting tools. An M06 is usually used for clamping or unclamping the work-holding device. The T-code or tool code is sufficient to tell the control which tool turret position to change to. The tool call is also accompanied by the offset call. The offset number is the last two digits in the tool call. For example, a T0101 would load tool 1 with an offset 1. These offsets are used for offsetting the tool path to accommodate for tool wear or exact sizing of the part. The programmer must be careful to return the axis to home or to a safe position with a G28 or G29 preparatory function before calling a new tool. If the tool turret is not in a safe position to index it could crash into the part or chuck and cause severe damage. Here is a look at a tool change block.

N0001 T0101;

The first two numbers call for tool #1, the second two numbers call for offset number 1. Normally, we correspond the tool offset number and the tool number to reduce mistakes in machining and damage to equipment.

Workpiece Coordinate Setting

Workpiece coordinate setting is easier on turning machines than on machining centers. The spindle or workpiece center is the location of the X0 position, and the Z0 position is typically the right end of the workpiece or the chuck face (see Figure 9–6).

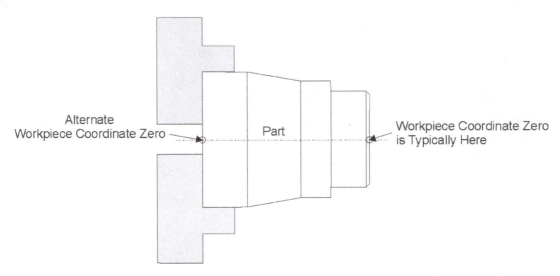

FIGURE 9–6. Possible workpiece coordinate location settings.

The workpiece location is set using an offset. Just as there are many different types of CNC controls, there are many different ways of using offsets to locate the workpiece.

G50 Method for Workpiece Offsets

When using the G50, the part datum location will accompany the code. For example, G50 X4.000 Z4.000; this would be the distance from the tool tip at the machine home position to the right end and center of the workpiece (see Figure 9–7).

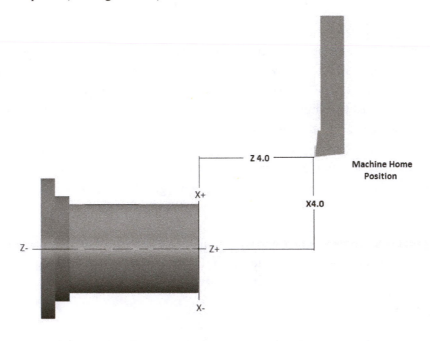

FIGURE 9–7. G50 tool offset. In this example, the code would be G50 X4.00 Z4.00.

Tool Geometry Method for Workpiece Offsets

The most common method of workpiece location used on turning centers is a tool geometry offset. The tool geometry offset is called or activated when the tool is called. When a line in the program calls T0101, tool number 1 indexes to the cutting position and the control looks in the offset page and loads offset 01.

Tool Offset X-Axis Setting

A piece of material must be put in the chuck before any tool setting or zero setting operations can be performed. Index the turret to the tool that is to be measured. Start the spindle using the manual mode. Make sure you are using the proper RPM. Use the turning tool to make a small cut on the diameter of the part (see Figure 9-8).

Figure 9-8. Making a cleanup cut on the diameter of the part.

Approach the part carefully and feed slowly during the cut. After the small cut is done, jog away from the part using only the Z-Axis. Move far enough away from the part so that you can take a measurement with your micrometer. Stop the spindle and open the door. Take a measurement on the turned diameter of the part. On newer controls there will be a button that records the machine position and automatically records this position in the offset table. On the HAAS control this is called the X DIAMETER MEASURE button (see Figure 9-9).

Figure 9-9. HAAS Control-X diameter measure button.

On the Okuma control this button is called the *CAL* or calculate button. If you have an older control you will have to locate the machine position screen and write down the X axis location (see Figure 9-10). Next you will have to add the measurement of the turned diameter to the X-axis machine location.

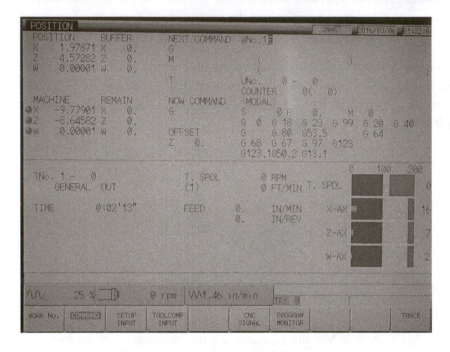

Figure 9-10. Machine position page.

On some controls this is done with an add button or the Enter button. If using a G54 offset, add the machine position to the measured diameter and enter this number in the G54 X position offset line of the program. The G54 X-axis offset is the distance from the machine home position to the tool tip when it is at the center of the spindle (See Figure 9-7). This procedure will need to be repeated for every tool used in your program.

Review:
- Take a cleanup cut on the diameter of the workpiece.
- Write down the X-axis machine position or press the appropriate X-axis measure button.
- Measure the diameter of the part.
- Add the diameter to the X-axis machine position.
- Put the offset (machine position + measured diameter) for this tool in the G50 command code line of your program.
- Repeat the procedure for every tool used in the program.

Tool Offset Z-Axis Setting

Index the turret to the tool that is to be measured. Make a small cut on the face of the material clamped in the spindle (see Figure 9-11). Approach the part carefully and feed slowly during the cut. After the small cut is done, jog away from the part using X-axis. Move far enough away from the part so that you can take a measurement with your measuring tool.

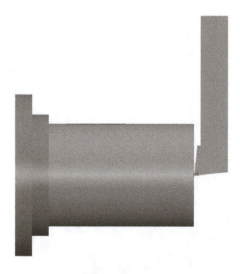

Figure 9-11. Taking a facing cut.

On newer controls there will be a button that records the machine position and automatically records this position in the offset table. On the HAAS control this button is called the *Z FACE MEASURE button* (see Figure 9-12).

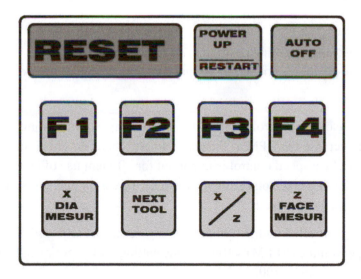

Figure 9-12. Haas Control buttons for setting tool offsets.

On the Okuma control this button is called the CAL or calculate button. If you have an older control you will have to locate the machine position screen and write down the Z axis location. If using a G54 offset, enter this number in the G54 Z position offset. The G54 Z-axis offset is the distance from the machine home position to the tool tip when it is at the Z0.0 location (face) of the part. Use a piece of paper or feeler gage to measure additional tools off the face of the part and add the feeler gage thickness to the calculation (see Figure 9-13).

Feeler Gage

Figure 9-13. Using Feeler Gage stock to touch off tools.

Review:

- Take a face cut
- Write down the Z-axis machine position
- Put this machine position in the G50 command code line of your program for this tool under the Z-axis.

Spindle Start Block

On the turning center, three codes control the spindle. A large majority of turning centers have the capability to increase or decrease the RPM as the part diameter changes. This is called *constant surface speed control*. Constant surface speed control is important for efficient use of cutting tools, tool life, and proper surface finish. As cutting takes place and the diameter being cut decreases, the spindle speed increases to ensure optimal cutting speeds.

Constant surface speed is controlled with a G96 preparatory code. The G96 code is followed by the proper surface footage per minute (SFPM) setting for the cutting tool material and part material. The value is set with an S: such as G96 S400.

When not using constant surface speed control, a G97 RPM input code is used. For thread cutting or drilling, a G97 code is used. The G97 will be followed by the properly calculated RPM: such as G97 S450. An M03 or M04 tells the spindle to start in a clockwise or counterclockwise direction. A typical spindle start block using constant surface speed control follows:

N0010 G96 S450 M03;

Block number 10 would set constant surface speed control to 450 SFPM and start the spindle in a clockwise direction.

Tool Motion Blocks

The tool motion blocks are the body of the program. The tool is positioned and the cutting takes place in these blocks.

Home Return

Tools need to be returned to home or a fixed position before a tool change (index) takes place. Most machine controls use a G28 command to rapid position the tool to home. On larger machines or for maximum efficiency, the operator can use a position closer to the chuck to change tools. When a home return is commanded, the path the tool takes to get there is crucial. If we use only a G28, the tool will take the straightest route to home.

A G28 enables the programmer to determine how it should move to home. If we wanted the tool to move up to a X2.00 position before returning home, we command G28 X2.00. This command would move the tool to X2.00 and then would automatically return home (see Figure 9–14). This can be especially useful when grooving because we want the tool to clear the workpiece before going to the home position. If the tool were inside a bore, we would command the tool to come out of the hole before rapid positioning to home. In most cases programmers like to use incremental axes commands, such as a U for X and a W for Z with the G28 command. This assures the direction of tool movement. In the case of a G28 U2.0 command, the tool would move up (incrementally) two inches in the X axis before returning both axes to home.

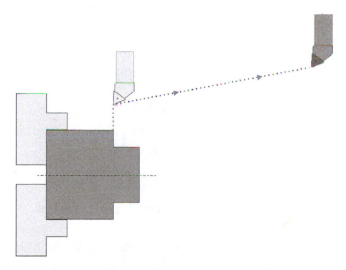

FIGURE 9–14. G28 return to home position command.

Program End Blocks

There are a number of ways to end a program. Some controls require that you turn off the coolant and the spindle with individual miscellaneous function codes. Other controls will end the program, reset the program, and turn off miscellaneous functions, with an M30 code. No matter what type of control you have, it is always a good idea to cancel any offsets that may be active when ending your program.

Next, we examine a typical turning program that incorporates many of the elements that have been covered. The part is shown in Figure 9–15.

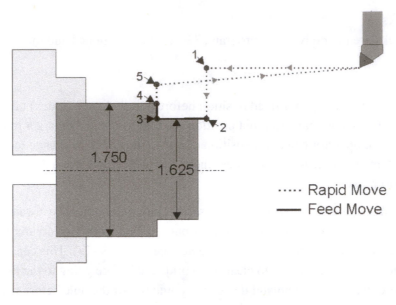

FIGURE 9–15. The center of the workpiece is X0.0 zero and the right end of the stock is Z0.0.

The program for the part shown in Figure 9–15 is listed below.

O0915 (Program name O0915)
N10 G90 G20 G40 (G90 - Absolute mode, G20 - Inch mode, G40 - Cancel tool nose radius compensation)
N15 G28 U0.0 W0.0 (Return to reference position - machine home in this example)
N20 G00 T0101 (Tool 1, offset 1)
N30 G50 X5.800 Z10.250 (Work offset for tool 1 at X5.8 Z10.25)
N40 G96 S400 M03 (Turn on spindle clockwise at 400 RPM constant surface speed)
N50 G50 S3600 (Set maximum spindle speed at 3600 RPM)
N60 G00 Z.1 (Rapid to Z.1)
N70 G00 X1.625 (Rapid to X1.625)
N80 G01 Z-1.25 F0.01 (Linear feed to Z-1.25 at a feed   rate of 0.01 per revolution)
N90 G01 X1.75 (Linear feed to X1.75)
N100 G28 X2.00 M05 (G28 - Return to reference point through X2.00, M05 - Spindle stop)
N110 T0100 (Cancel the offset for tool 1)
N120 M30 (Program end/memory reset)
%

Circular Interpolation

Only straight-line moves were used in the program to machine the part in Figure 9–15. One of the most important capabilities of a CNC machine is its ability to do circular cutting motions. CNC turning centers are capable of cutting an arc of a specified radius value. Arc, or radius, cutting is known as circular interpolation.

Circular interpolation is done in the same manner as on machining centers, with the use of G02 or G03 preparatory codes. To cut an arc, the programmer needs to follow a specific procedure. To cut an arc, the tool needs to be positioned to the start point of the arc in the line before the circular interpolation will be done. We need to tell the control the direction of the arc in the circular interpolation line: clockwise or counterclockwise. The third piece of information is the end point of the arc. The last piece of information the control needs is the position of the arc center or, if you are using the radius method of circular interpolation, the radius value of the arc (see Figure 9–16).

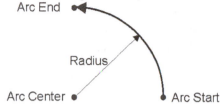

FIGURE 9–16. The critical pieces of information needed to cut an arc are the arc start point, arc direction, arc end point, and arc centerpoint location.

Arc Start Point

The arc start point is the coordinate location of the start point of the arc. The tool is moved to the arc start point in the line prior to the arc generation line. Simply stated, the start point of the arc is the point the tool is presently at when you want to begin cutting the arc.

Arc Direction (G02, G03)

Circular interpolation can be carried out in two directions: clockwise or counterclockwise. Two G-codes specify arc direction. The G02 code is used for circular interpolation in a clockwise direction, and the G03 code is used for circular interpolation in a counterclockwise direction. Both codes are modal. G02 and G03 codes are controlled by a feed rate (F) code, just like a G01.

Arc End Point

The tool must be positioned at the start point of the arc prior to a G02 or G03 command. The current tool position becomes the arc starting point and the arc end point is the coordinate position for the end point of the arc. The arc start point and arc end point determine the tool path, which is generated according to the arc center position.

Arc Centerpoints

To generate an arc path, the controller has to know where the center of the arc is. There are two methods of specifying arc centerpoints: the coordinate arc center-point method and the radius method.

When using the coordinate arc center method, a particular problem arises: How do we describe the position of the arc center? If we use the traditional X, Y, Z coordinate position words to describe the end point of the arc, how will the controller discriminate between the end points coordinates and the arc center coordinates? It is done by using different letters to describe the same axes. Secondary axes addresses are used to designate arc centerpoints. The secondary axes addresses for the primary axes are:

I = X axis coordinate of an arc centerpoint

K = Z axis coordinate of an arc centerpoint

When we cut an arc on the turning center, the X/Z axes are the primary axes, and the I/K letter addresses are used to describe the arc centerpoint. The type of controller that you use determines how these secondary axes are located.

With some controllers, such as the HAAS or Fanuc controller, the arc centerpoint position is described as the incremental distance from the arc start point to the arc center. This is the most common method of specifying the arc center.

Some CNC controllers can calculate the centerpoint of the arc by merely stating the arc size (radius) and the end point of the arc. We will concentrate on the incremental method of defining the arc center. Keep in mind that we are locating the arc center using the incremental method.

If the arc centerpoint is located down or to the left of the start point, a negative sign (–) must precede the coordinate dimension.

Most CNC turning center controllers are programmed using diameter values, so if we move the tool out 1/4 of an inch, we change the diameter by 1/2 of an inch. If we have an arc of a .25 inch radius, the diameter of the part can change by .50 of an inch.

Study the example in Figure 9-17. The line of code is G03 X.500 Z-.250 I0.0 K-.250. Note that the X value end point was a diameter value. Note also that the I and K represented the distance from the start to the center of the arc in the X and Z directions.

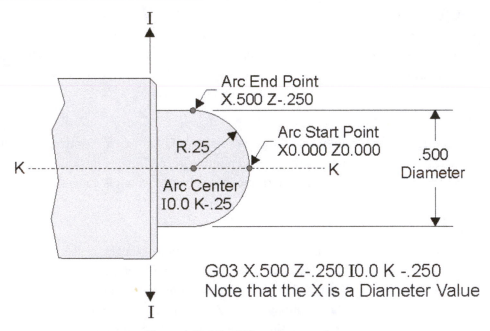

FIGURE 9–17. G03 arc generation.

The next programming example (see Figure 9–18) uses circular interpolation. Pay particular attention to the locations of the start, end, and center point locations.

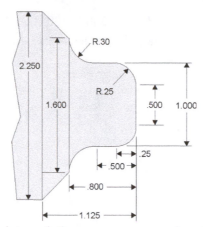

FIGURE 9–18. Arc interpolation program example using I and K values.

O0918 (Program name O0918)

N10 G90 G20 G40 (G90 - Absolute, G20 - Inch, G40 - Cancel tool nose-radius compensation)

N15 G28 U0.0 W0.0 (Return to reference position - machine home in this example, this makes sure that the machine is in the correct position before cutting)

N20 G00 T0101 (Tool 1, offset 1)

N30 G50 X5.800 Z10.250 (G50 - Work offset for tool 1 X5.800 Z10.25)

N40 G96 S400 M03 (Spindle on clockwise at 400 surface feet per minute constant surface speed)

N50 G50 S3600 (Set maximum spindle speed at 3600 RPM)

N60 G00 Z.1 (Rapid to Z.1)

N70 G00 X0.0 (Rapid to X0.0)

N80 G01 Z0.0 F0.008 (Linear feed to Z0.0 at a feed rate of .008 inches per revolution)

N90 G01 X.50 (Linear feed to X.50)

N100 G03 X1.00 Z-.25 I0.0 K-.25 (Counterclockwise circular interpolation)

N110 G01 Z-.50 (Linear feed to Z-.50)

N120 G02 X1.60 Z-.80 I.30 K0.0 (Clockwise circular interpolation - IJ method)

N130 G01 X2.25 Z-1.125 (Linear feed to X2.25 Z-1.125)

N140 G28 X3.00 M05 (G28 - Return to the reference point through X3.00, M05 - spindle stop)

N150 T0100 (Cancel tool offset values in tool 1)

N160 M30 (Program end/memory reset)

%

Radius Programming Method for Circular Interpolation

The direct arc radius programming for arc interpolation is easier to use than the traditional I and K method. To use direct radius programming method follow these steps:

- Position the tool to the start point of the arc.
- Define the direction of the arc. Arc direction can be clockwise (G02) or counterclockwise (G03).
- Define the end point of the arc in the X and Z axes.
- Define the radius value of the arc.

Study the examples shown in Figure 9–19. The radius on the left was cut with G03 (counterclockwise circular interpolation). The one on the right was cut with G02 circular interpolation. Both utilized the radius method to specify the center of the arc.

The code for the one on the left was G03 X2.000 Z-.500 R.500. The R.500 in the code specified that the radius was .500. This information plus the endpoint values is enough for the machine to cut this arc.

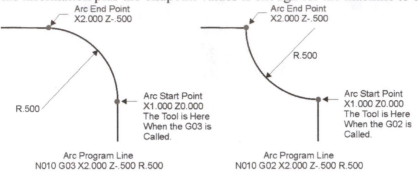

FIGURE 9–19. Arc programming.

Figure 9-20 shows a sample turned part. The program for it follows. The program uses radius values to turn the arcs. Study the program.

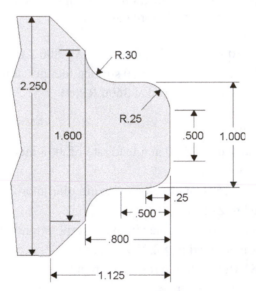

FIGURE 9–20. Circular interpolation program example using radius programming.

O009 (Program name – O009)

N10 G90 G20 G40 (G90 - Absolute mode, G20 - Inch mode, G40 - Cancel tool nose-radius compensation)

N15 G28 U0.0 W0.0 (Return to reference position - machine home in this example, this makes sure that the machine is in the correct position before cutting)

N20 G00 T0101 (Tool 1, offset 1)

N30 G50 X5.800 Z10.250 (G50 - Work offset for tool 1 - X5.800 Z10.25)

N40 G96 S400 M03 (Spindle on clockwise at 400 surface feet per minute constant surface speed)

N50 G50 S3600 (Set maximum spindle speed at 3600 RPM)

N60 G00 Z.1 (Rapid to Z.1)

N70 G00 X0.0 (Rapid to X0.0)

N65 G01 Z0.0 F.008 (Linear feed to Z0.0 at a feed rate of .008 inches per revolution)

N80 G01 X.50 F0.008 (Linear feed to X.50)

N90 G03 X1.00 Z-.25 R.25 (Counterclockwise circular interpolation using the radius method)

N91 G01 Z-.50 (Linear feed to Z-.50)

N92 G02 X1.60 Z-.80 R.30 (Clockwise circular interpolation - radius method)

N93 G01 X2.25 Z-1.125 (Linear feed to X2.25 Z-1.125)

N100 G28 X3.00 M05 (G28 - Return to the reference point through X3.00, M05 - Spindle stop)

N110 T0100 (Cancel tool offset values in tool 1)

N120 M30 (Program end/memory reset)

%

Tool-Nose Radius (TNR) Compensation

Tool-nose radius compensation is a type of offset used to control the shape of machined features. The turning center control can offset the path of the tool so that the part can be programmed just as it appears on the part print. There is an error that becomes apparent when we use the tool edges to set our workpiece coordinate position on a turning center. When we set the X and Z axes of the tool, we specify a single sharp point. Most of the tools we use for turning have radii. These radii on the tool nose can create inaccuracies in the cutter path (see Figure 9-21). The dotted lines represent the path the tool would take. The solid line represents the desired part shape.

You can see that the part shape would be incorrect if tool-nose radius compensation is not used. The cutting point on the tool changes as its orientation to the part path changes, leaving inaccuracies. To compensate for the radii of the cutting tool, tool-nose radius compensation must be used, which saves us from having to mathematically calculate a cutter path that would compensate for the radius of the tool.

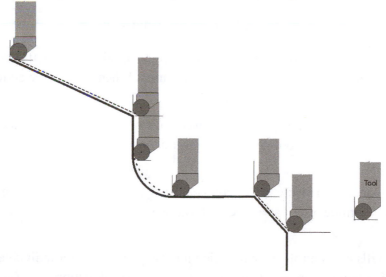

FIGURE 9–21. The dotted lines above show the errors that could occur due to the radius of the cutting tool not being a sharp point. Tool-nose radius compensation allows us to program the part, not the tool path.

TNR compensation also lets us use the same program for a variety of tool types. With TNR compensation capabilities, the insert radius size can be ignored and the part profile can be programmed. The exact size of the cutting tool to be used is entered into the offset file, and when the offset is called, the tool path will automatically be offset by the tool radius.

Tool-nose radius compensation can be to the right or left of the part profile. To determine which offset you need, imagine yourself walking on the programmed path in the direction of the cut. Do you want the tool to the left of the programmed path or to the right (see Figure 9–22)

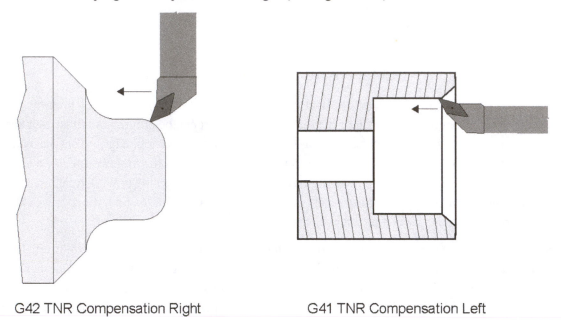

G42 TNR Compensation Right   G41 TNR Compensation Left

FIGURE 9–22. Tool-nose compensation.

Compensation direction is controlled by a G-code. When compensation to the left is desired, a G41 is used. When compensation to the right is needed, a G42 is used. When using these cutter compensation codes, you need to specify how large the offset needs to be.

The size of the radius is placed in the nose radius offset table, which is typically located in the tool file under the tool number being used. Tool-nose radius information can be determined from catalogs or the insert package.

The other information needed to insure proper compensation is the tool-nose direction vector. The tool tip or imaginary tool tip of turning tools has a specific location or direction from the center of the tool-nose radius.

The tool-nose vector tells the control which direction it must compensate for individual types of tools. Standard tool-nose direction vectors are shown in Figure 9–23. The direction vector number is usually placed in the same tool offset table as the radius value.

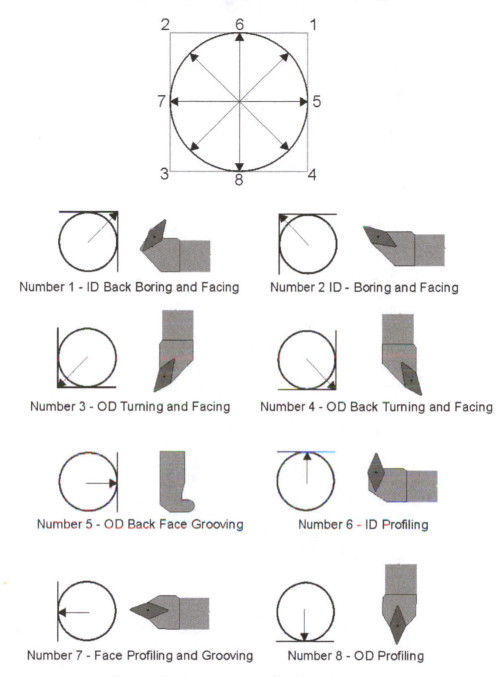

FIGURE 9–23. Tool-nose radius direction vectors.

Lead In and Lead Out Moves Associated with Tool Nose Radius Compensation

The first linear (G00 or G01) move in a line that contains a G41 or G42 is called a lead in or approach move. This first move is not compensated, but by the end of the move the tool nose radius should be fully compensated for. The distance of the lead in move should be more than the tool nose radius compensation. Lead in moves should start away from the part (see Figure 9-24).

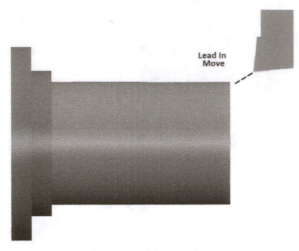

Figure 9-24. Lead in move.

A G40 line of code cancels the tool nose radius compensation. The linear line move that is associated with the G40 is called a lead out or departure move (see Figure 9-25). The tool nose radius at the start of the departure move is compensated, but by the time the tool moves to the destination the tool nose radius is no longer being compensated. The distance of the lead out move should be more than the tool nose radius compensation. A G40 should be called when the tool is clear of the part.

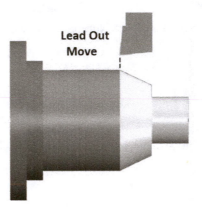

Figure 9-25. Lead out move.

Figure 9–26 illustrates a typical part that uses tool-nose radius compensation. The program for the part follows the figure.

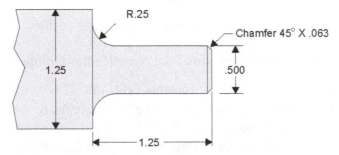

FIGURE 9–26. Part that will utilize tool-nose radius compensation in the program.

O0915 (Program name O0915)

N5 G90 G20 G40 (G90 - Absolute mode, G20 - Inch mode, G40 - Cancel tool nose-radius compensation)

N10 G28 U0.0 W0.0 (Return to reference position - machine home in this example)

N15 G00 T0101 (Tool 1, offset 1)

N20 G50 X5.80 Z10.25 (G50 - Work Offset for tool 1 at X5.800 Z10.25)

N25 G0 Z.1 M8

N30 G50 S3600 (Set maximum spindle speed at 3600 RPM)

N35 G96 S400 M03 (Spindle on clockwise at 400 surface feet per minute constant surface speed)

N40 G1 G42 Z0. F.01 (Linear move, tool-nose radius compensation on the right side, lead in move to Z0. at a feed rate of .01 inches per revolution)

N45 X.5 Z-.0625 (Linear move to X.5 Z-.0625)

N50 Z-1. (Linear move to Z-1.)

N55 G2 X1. Z-1.25 R.25 (Clockwise circular move)

N60 G1 X1.25 (Linear move to X1.25)

N65 G40 X1.45 (Tool-nose radius compensation cancel, lead out move)

N70 M9 (Coolant off)

N75 G28 U0. W0. M05 (G28 - Return to the reference point through incremental U0.0 W0.0 M05 - Spindle stop)

N80 T0100 (Cancel tool offset values in tool 1)

N85 M30 (Program end/Memory reset)

%

Chapter Questions

1. Which turning center axis controls the diameter of the part?
2. Define constant surface footage control.
3. What secondary axes addresses are used to define the centerpoint position when cutting radii on a turning center?
4. Explain why tool-nose radius compensation is needed when cutting tapers or radii.
5. Use the part print to fill in the point locations.

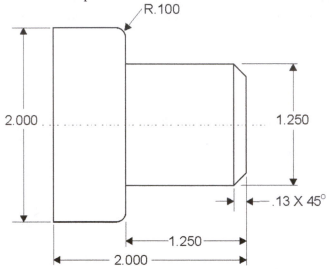

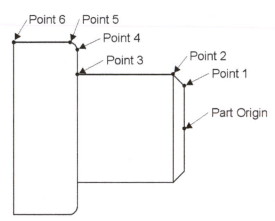

Point #	X (Diameter) Value	Z Value
Point 1		
Point 2		
Point 3		
Point 4		
Point 5		
Point 6		

6. Use the part print to fill in the point locations. Part origin is located at the right end and center of the part.

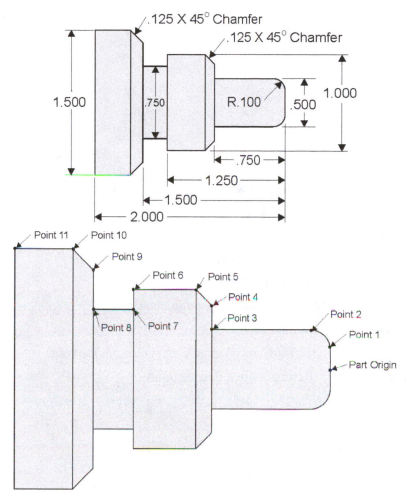

Point #	X (Diameter) Value	Z Value
Point 1		
Point 2		
Point 3		
Point 4		
Point 5		
Point 6		
Point 7		
Point 8		
Point 9		
Point 10		
Point 11		

7. Fill in the blanks to complete the program to machine the profile of the part in question 5.

```
O0008
N10 G90 G20 G40
N15 G28 U0.0 W0.0
N20 G00 T0101
N25 G50 X10.00 Z12.00
N30 G00 Z.1
N35 G00 X.99
N40 G50 S3600
N45 G96 S200 M___          Spindle Forward
N50 G01 G__ Z0. F.01    Tool Nose Radius Compensation Right, lead in to (POINT 1)
N55 G01 X___ Z-.13      X Coordinate at (POINT 2)
N60 G01 Z____           Z Coordinate at (POINT 3)
N65 G01 X___        X Coordinate at (POINT 4)
N70 G__ X2__ Z__ K-.1 CCW arc to X and Z Coordinate at (POINT 5)
N75 G01 Z___            Z Coordinate at (POINT 6)
N80 G01 G__             Cancel Tool Nose Radius Compensation, lead out
N85 G__ U0. W0. M05        Return to the reference point
N90 T0100
N95 M30
%
```

8. Number the procedures in the proper order.

_____ Tool call

_____ Start-up procedures

_____ Workpiece location block

_____ Home return

_____ Spindle speed control

_____ Tool motion blocks

_____ Program end procedures

9. Use the part print to fill in the point locations. Part Origin is located at the right end and center of the part.

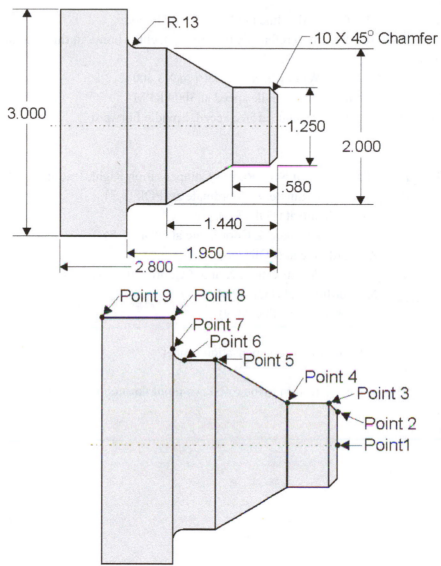

Point #	X (Diameter) Value	Z Value
Point 1		
Point 2		
Point 3		
Point 4		
Point 5		
Point 6		
Point 7		
Point 8		
Point 9		

10. Fill in the blanks to complete the program to machine the profile of the part in question 9.

```
O0009
N10 G___ G ____ G40    G90 - Absolute mode, G20 - Inch mode
N15 G ___ U.0.0 W0.0    Return to reference position - machine home in this example
N20 G00 T0101
N25 G___ X _____ Z _____    Work offset for tool 1 at X6.800 Z12.25
N30 G__ S3000    Set maximum spindle speed at 3000 RPM
N35 G___ S200 M____    Set constant surface speed, Spindle Forward
N40 G00 Z.1
N45 G00 X.1.05
N50 G01 G___ Z ___ F.01    Tool Nose Radius Compensation Right, lead in to (POINT 2)
N55 G01 X___ Z ____    X Coordinate, Z Coordinate at (POINT 3)
N60 G01 Z___    Z Coordinate at (POINT 4)
N65 G01 X___ Z ____    X Coordinate, Z Coordinate at (POINT 5)
N70 G01 Z___    Z Coordinate at (POINT 6)
N75 G __ X ___ Z ___ R __    Arc direction, X and Z Coordinate at (POINT 7), Arc radius
N80 G __ X ___    X Coordinate at (POINT 8)
N85 G01 Z ___    Z Coordinate at (POINT 9)
N90 G01 X3.10
N95 G__ X3.25    Tool-nose radius compensation cancel
N100 M___    Coolant off
N105 G__ U ___ W___ M05    Return to the reference point through incremental X0.0 Z0.0
N110 T0100
N115 M30
%
```

Chapter 10

Programming Canned Cycles for CNC Turning Centers

INTRODUCTION

This chapter will examine canned cycles for turning centers. There are canned cycles for many types of operations. There are canned cycles to simplify roughing out a part, finish machining, tapping, threading, boring, drilling and deep hole drilling, and many other common operations. Canned cycles can dramatically simplify programming. Note: you should check the manual for your particular control before using any canned cycles to be sure they are programmed in the same manner. There are differences between different machine controls.

OBJECTIVES

Upon completion of this chapter, the reader will be able to:

- Explain the use of various canned cycles.
- Write programs using roughing and finishing canned cycles.
- Write programs using drill canned cycles.
- Write programs using grooving canned cycles.
- Write programs using boring canned cycles.

Canned Cycles for Turning Centers

Canned cycles (fixed) cycles are used to simplify the programming of repetitive turning operations, such as rough turning, threading, and grooving. Canned cycles are sets of preprogrammed instructions that can dramatically shorten the number of lines in a program. Programming a simple part without the use of a canned cycle can take up to four or five times the number of lines needed for a part programmed with canned cycles. Think of the operations that are needed to produce a thread:

     1 - Position the X and Z axes to the proper coordinates with a rapid traverse move (G00).

     2 - Position the tool for the proper lead angle.

     3 - Feed the tool across.

     4 - Rapid position the tool back to the clearance plan.

     5 - Feed the tool across, and repeat this many times increasing the depth of cut each time.

By using a canned threading cycle, a thread can be done with one line of programming. See Figure 10–1 for a general list of the most commonly used canned cycles for turning centers.

Fixed (canned cycles) may vary between CNC machines. Although the cycles work in the same basic manner, always refer to the programming manual for your specific machine when trying a new canned cycle.

Common Canned Cycles for Turning Centers	
G70	Finishing Cycle
G71	Roughing Cycle
G72	Face Stock Removal
G76	Thread Cycle- Multiple Pass
G81	Drill Canned Cycle
G82	Spot Drill/Counterbore
G83	Peck-Deep Hole Drilling
G84	Tapping Canned Cycle
G85	Bore In-Bore Out Cycle
G86	Bore in-Stop-Rapid Out Cycle

Figure 10-1. Common turning center canned cycles.

Roughing Cycle (G71)

The G71 automatically takes roughing passes to turn down a workpiece to the required part profile at a specified depth of cut. The G71 roughing cycle reads a specified number of program blocks to determine the part profile, depth of cut, and feed rate. From the parameters in the program the control then calculates all of the moves to rough the part out from the rough stock. Cutting is accomplished through parallel moves of the tool in the Z-axis direction.

Study Figure 10-2. This figure shows the rough stock and the finished part. Without canned cycles the programmer would need to program each of the passes to make this part. This would be quite a bit of programming. With a roughing canned cycle the operator simply defines the coordinates of the finished part in the program, the depth of cut to be used and a feed rate. Then the operator programs a rapid move to position the cutting tool to just outside the corner of the rough stock shape before the roughing canned cycle is called. The roughing canned cycle then calculates all of the moves necessary to take several roughing cuts to machine the part. A finishing canned cycle can then be called to use the same part coordinates to calculate the moves to make a finish pass to complete the part.

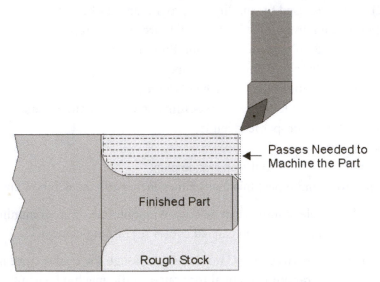

Figure 10-2. Machining passes needed to rough out a part.

Certain steps need to be followed when using canned cycles. In the first step, the tool needs to be positioned to the rough stock boundaries. This step has a dual purpose: it tells the control how large the stock is, and it creates a Z clearance position that the tool rapids back to for each pass. The G71 uses letters to give the controller information on the part profile, the amount of stock to leave for finishing, the depth of cut, and the feed rate. A G71 roughing cycle command is shown in Figure 10-3.

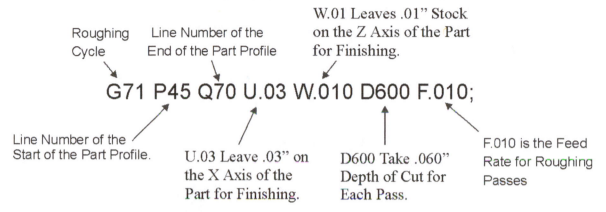

Figure 10-3. Example of a G71 roughing cycle call.

G71 is the roughing cycle call.

P45 is the block or line number that designates the start of the part profile.

Q70 is the block or line number that designates the end of the part profile.

U.03 tells the controller that we want to leave .03 of an inch stock on the X axis (diameter) of the profile for finishing.

W.010 tells the controller that we want to leave .01 of an inch stock on the Z axis (length) of the profile for finishing.

D.060 tells the controller to take .060" depth of cut for each pass. This has become the most common way to specify the depth of cut. *Some controls are different however. It is very important that you check the manual for your machine. One some controls the D value does not use a decimal point. On these controls the D is commanded the controller reads from the right and decides what the depth of cut will be. Each number in each decimal position gets a value. If we wanted to take .050" depth of cut per side, we would write it as D500. The first zero from the right has 0 tenths of a thousandths value. The next zero from the right has 0 thousandths of an inch value. The 5 in the third position from the right has 5 ten-thousandths of an inch value or 50 thousandths of an inch. Make sure you check on the correct format for your machine control.*

F.010 is the feed rate of the roughing passes.

Figure 10-4 shows a sample part that would be appropriate to make using a roughing canned cycle. Figure 10-5 shows an example of the cutting passes that a roughing canned cycle might make.

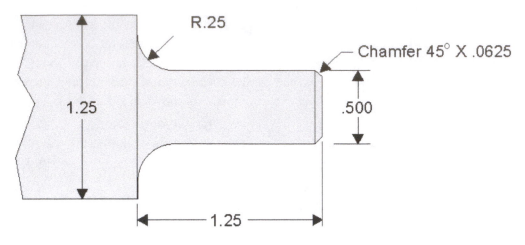

Figure 10-4. Sample part.

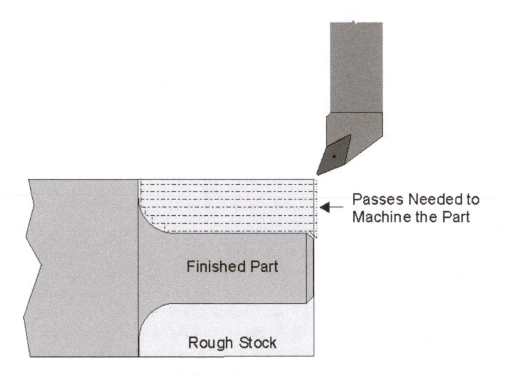

FIGURE 10–5. Roughing cycle example for the part shown in 10-4.

Figure 10-6 shows the coordinates of the finished part shape that need to be defined for the G71 roughing or G70 finishing caned cycle. Note that they are defined in lines N45 tough N70 in the program that follows. In the canned cycle call P45 will point to the first line in the program that defines the start point of the part shape and Q70 will point to the line in the program that has the coordinates of the end point of the part shape.

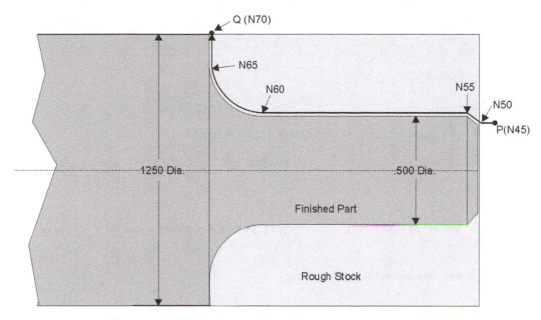

Figure 10-6. Definition of the finished shape of the part.

A program that utilizes a G71 roughing cycle to make the part shown in Figure 10-4 follows. TNR compensation will not be used in this example.

O0916 (Program name O0916)

N5 G90 G20 G40 (G90 - Absolute mode, G20 - Inch mode, G40 - Cancel tool- nose radius compensation)

N10 G28 U0.0 W0.0 (Return to reference position - machine home in this example)

N15 G00 T0101 (Tool 1, offset 1)

N20 G50 X5.80 Z10.25 (G50 - Work offset for tool 1 at X5.800 Z10.25)

N25 G50 S3600 (Set maximum spindle speed at 3600 RPM)

N30 G96 S400 M03 (Spindle on clockwise at 400 surface feet per minute constant surface speed)

N35 G00 X1.30 Z.100 (Rapid to X1.30 Z.100)

N40 G71 P45 Q70 U.03 W.01 D.060 F.01 (G71 - Roughing cycle, P45 - N45 first line of part profile, Q70 - last line of part profile, U - .03 Leave .03 on the X axis for finish cut, W.01- Leave .010 on the Z axis for finish cut, D.060 –depth of cut, radius value. *(NOTE: some machines do not use a decimal point for the depth of cut value. Make sure you check your machine so that you format the depth of cut correctly, F.01 - feed of .01 inches per revolution)*

N45 G00 X.375 M8 (G00 - Rapid to X.375, M8 - coolant on)

N50 G1 Z0 F.01 (Linear move to Z0. at a feed rate of .01 inches per revolution)

N55 G1 X.5 Z-.0625 (Linear move to X.5 Z-.0625)

N60 G1 Z-1.0 (Linear move to Z-1.)

N65 G2 X1.0 Z-1.25 R.25 (Clockwise circular move)

N70 G1 X1.25 (Linear move to X1.25)

N75 M9 (Coolant off)

N80 G28 U2.0 M05 (Return to reference though U2.0 incremental)

N85 T0100 (Note that this is done to cancel the offset for tool 1. Note also that on some controls this is unnecessary. A tool call in the next line would cancel the tool 1 offset and implement the tool 2 offset.)

N90 G28 U2.0 M05 (G28 - Return to the reference point through incremental U2.0, M05 - spindle stop)

N95 M30 (Program end/Memory reset)

%

Figure 10-7 shows the parameters that can be used for a G71 roughing canned cycle.

G71 Parameters	
P	Block number of the start of the part shape
Q	Block number of the end of the part shape
U*	Finish stock remaining with direction (+or -), X-axis diameter value
W*	Finish stock remaining with direction (+or -), Z-axis value
I*	Last pass amount with direction (+or -), X-axis radius value
K*	Last pass amount with direction (+or -), Z-axis value
D*	Depth of cut stock removal each pass, positive radius value (HAAS Setting 72)
F	Feed rate
S**	Spindle speed in this cycle
T**	Tool and offset in this cycle
* Optional ** Rarely defined in a G71 Line	

Figure 10-7. Parameters for a G71.

Finish Cycle (G70)

The G70 command determines the finish part profile, and then executes a finish pass on the part. The finishing cycle is called with a G70, followed by a letter address P for the start line of the finish part profile and the letter address Q for the end line of the part profile (see Figure 10-8). Note that the part profile would normally already have been specified by a roughing cycle. The finishing canned cycle could use the same part profile information that was already defined in the program.

A finish feed rate can also be included in the finishing canned cycle block. When the finish cycle is commanded, it reads the program blocks designated by the P and Q and formulates a finishing cycle based on the lines between P and Q that specify the part shape.

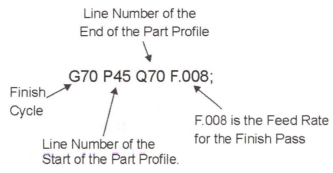

Figure 10-8. A finishing canned cycle block.

As with the G71 roughing cycle, the tool needs to be positioned to a Z clearance plane or stock boundary prior to the calling of the G70 finish cycle. The program below performs the roughing and finish canned cycles to machine the part shown in Figure 10–4. Also note that the roughing cycle and finish canned cycles both utilize the same part shape specified in program lines N45 (P) though N70 (Q).

O0916 (Program name O0916)

N5 G90 G20 G40 (G90 - Absolute mode, G20 - Inch mode, G40 - Cancel tool- nose radius compensation)

N10 G28 U0.0 W0.0 (Return to reference position - machine home in this example)

N15 G00 T0101 (Tool 1, offset 1)

N20 G50 X5.80 Z10.25 (G50 - Work Offset for tool 1 at X5.800 Z10.25)

N25 G50 S3600 (Set maximum spindle speed at 3600 RPM)

N30 G96 S400 M03 (Spindle on clockwise at 400 surface feet per minute constant surface speed)

N35 G00 X1.30 Z.100 (Rapid to X1.30 Z.100)

N40 G71 P45 Q70 U.03 W.01 D.06 F.01 (G71 - Roughing cycle, P45 - N45 first line of part profile, Q70 - last line of part profile, U.03 -leave .03 on the X axis for finish cut, D.06 -depth of cut, F.01 - feed of .01 inches per revolution)

N45 G00 X.375 M8 (G00 - Rapid to X.375, M8 - coolant on)

N50 G1 Z0. F.01 (Linear move to Z0.0 at a feed rate of .01 inches per revolution)

N55 G1 X. 5 Z-.0625 (Linear move to X.5 Z-.0625)

N60 G1 Z-1. (Linear move to Z-1.)

N65 G2 X1. Z-1.25 R.25 (Clockwise circular move)

N70 G1 X1.25 (Linear move to X1.25)

N75 M9 (Coolant off)

N80 G28 U2.0 M05 (Return to reference though U2.0 incremental)

N85 T0100 (Note that this is done to cancel the offset for tool 1. Note also that on some controls this is unnecessary. The tool call in the next line would cancel the tool 1 offset and implement the tool 2 offset.)

N90 T0202 (Tool 2, Offset 2)

N95 G50 X6.75 Z9.625 (G50 - Tool offset for tool 2 at X6.750 Z9.625)

N100 G50 S3600 (Set maximum spindle speed at 3600 RPM)

N105 G96 S600 M03 (Spindle on clockwise at 600 surface feet per minute constant surface speed)

N110 G00 X1.30 Z.100 (Rapid move to X1.30 Z.100)

N115 G70 P45 Q70 F.008

N120 G28 U2.0 M05 (G28 - Return to the reference point through incremental U2.0, M05 - spindle stop)

N125 T0200 (Cancel tool offset values in tool 2)

N130 M30 (Program end/Memory reset)

%

Internal Roughing and Finishing Canned Cycles (G71/G70)

Figure 10-9 shows a part that would require a roughing and finishing cycle.

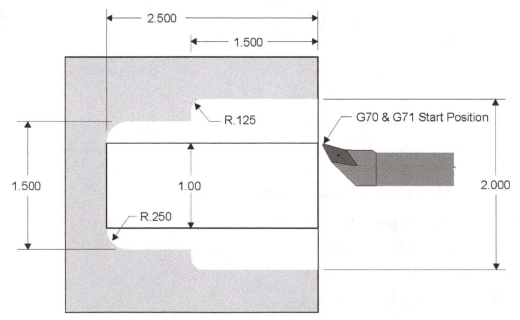

Figure 10-9. Example of internal roughing and finishing.

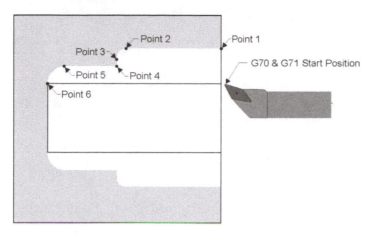

Figure 10-10. Points that establish the finished shape.

```
O7071 (Program name O0916)
N5 G90 G20 G40
N10 G28 U0.0 W0.0
N15 G00 T0101
N25 G50 S3600
N30 G96 S400 M03
N35 G00 X1.00 Z.100
N40 G71 P45 Q85 U-.03 W.01 D.030 F.01 (Same as external except U-.03 (negative) leaves stock
for finish boring)
N45 G00 X2.00 M8
N50 G01 Z0 F.01
N55 G01 X2.0 Z-1.375
N60 G03 X1.75 Z-1.50 R.125
N65 G01 X1.50
N80 G01 Z-2.25
N85 G03 X1.0 Z-2.5 R.25
N90 M9
N95 G28 U0.0 W1.0 M05
N100 T0202
N115 G50 S3600
N120 G96 S500 M03
N125 G00 X1.00 Z.100
N130 G70 P45 Q85 F.008
N135 G28 U0.0 W1.0 M05
N140 T0200
N145 M30
%
```

Drill Cycle (G81)

This cycle can be used to drill holes. Study Figure 10-11. The Z parameter is used to tell the machine how deep to drill. The R parameter tells the machine to rapid to the specified R Plane distance (see Figures 10-11 and 10-12). This cycle rapids to the R plane and then feeds at the commanded feed rate to the depth specified by the Z parameter. Once the drill reaches depth it rapids back out to the R plane.

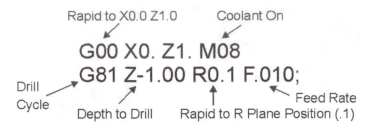

Figure 10-11. A G81 drill cycle.

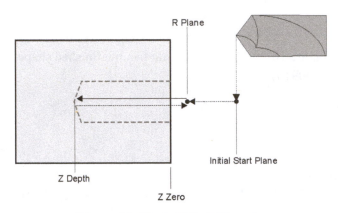

Figure 10-12. A G81 drill cycle.

Deep Hole Drill Cycle (G83)

This cycle can be used to drill deep holes. Study Figure 10-13. The Z parameter is used to tell the machine how deep to drill.

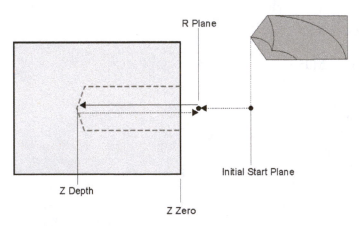

N050 G83 Z-1.625 R0.1 Q0.15 F0.005

Figure 10-13. Example of a G83 canned cycle.

A program using a G83 follows. The parameters for a G83 are shown in Figure 10-14.

O00120 (Program 120)
N010 G28

N020 T0101 (5/8 Spot drill) (Tool 1 - Offset 1)
N030 G97 S1450 M03
N040 G00 X0.0 Z1.0 M08 (Rapid to Initial Start Point)
N050 G83 Z-1.625 R0.1 Q0.15 F0.005 (G83 Deep Hole Drill with a full retract to R plane, Q is incremental depth of cut before full retract)
N060 G80 G00 Z1. M09
N070 G28 U0.0 W1.0
N080 M30
%

G83 Parameters	
X*	Absolute X axis rapid location
Z*	Absolute Z depth
Q*	Pecking depth amount
W*	Z Axis incremental pecking
P	Dwell time at Z-depth
R	R Plane
F	Feed rate
* Optional	

Figure 10-14. G83 canned cycle Parameters.

Peck Drilling Cycle (G74)

The G74 peck drilling cycle will peck drill holes with automatic retract and incremental depth of cut. The G74 command specifies the incremental depth of cut, the full depth of the hole, and the feed rate through the command variables K, Z, and F. Figure 10-15 shows an example of the proper format for peck drilling.

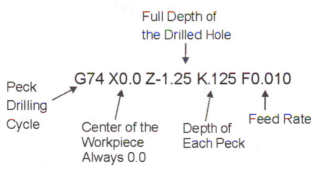

Figure 10-15. Example of a G74 canned cycle call.

G74 is the peck drill canned cycle.

X0.0 is the center of the workpiece (X is always zero).

Z-1.25 is the full depth of the drilled hole.

K.125 is the depth of each peck.

F.01 is the drilling feed rate.

The drill must be positioned to a clearance plane in the Z axis and also to X0.0 prior to the calling of the G74 peck drilling cycle. The spindle should also be reprogrammed for direct RPM using a G97 when drilling. Examine the sample peck drilling cycle in Figure 10–16.

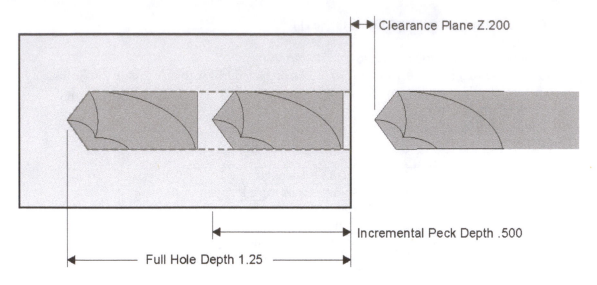

Figure 10–16. Peck drilling cycle.

O0917 (Program name O0917)

N10 G90 G20 G40 (G90 - Absolute mode, G20 - Inch mode, G40 - Cancel tool- nose radius compensation)

N20 G28 U0.0 W0.0 (Return to reference position - machine home in this example)

N30 G00 T0606 (Tool 6, offset 6)

N40 G97 S800 M03 (G97 - Spindle on clockwise at 800 RPM)

N50 G00 Z.2 (Rapid to Z.2)

N60 G00 X0.0 (Rapid to X0.0)

N70 G74 X0.0 Z-1.25 K.500 F0.01 (Peck drilling cycle)

N80 G28 Z1.00 M05 (G28 - Return to the reference point through Z1.00, M05 - spindle stop)

N90 T0600 (Cancel tool offset values in tool 6)

N100 M30 (Program end/Memory reset)

%

A G74 canned cycle can also be used for grooving on the face of a part (see Figure 10-17), turning with a chip break or high speed peck drilling. Figure 10-18 shows the parameters for a G74 canned cycle.

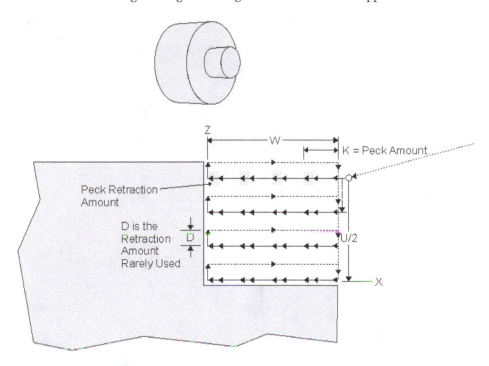

Figure 10–17. A G74 canned cycle being used to groove the face of a part.

G74 Parameters	
X*	Absolute X location to the furthest peck as a diameter value
Z	Absolute Z pecking depth
U*	X Axis incremental distance (+ or -) to the furthest peck, diameter value
W*	Z Axis incremental pecking depth
I*	X-Axis shift increment between pecking cycles positive radius value
K*	Z Axis pecking depth increment
D*	Tool shift amount when returning to the clearance plane. Note that the D value is rarely used.
F	Feed rate
* Optional	

Figure 10–18. Parameters for a G74 canned cycle.

Grooving Cycle (G75)

The grooving cycle is a very versatile canned cycle. To use the grooving cycle, the tool must be positioned to the start of the groove prior to calling the grooving cycle. Through a series of letter addresses, the controller can be commanded to cut a groove of varying width and depth.

Figure 10-19 shows an example of a grooving cycle. This example will cut one groove. Note that the groove is wider than the tool. The grooving tool is .25" wide. The groove is .375" wide. Figure 10-20 shows the shows the G75 line of code that would cut this groove.

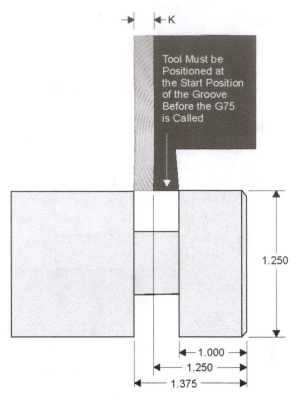

K

Tool Must be Positioned at the Start Position of the Groove Before the G75 is Called

1.250

1.000
1.250
1.375

FIGURE 10–19. This figure shows a part that would be appropriate for a G75 multiple pass grooving cycle.

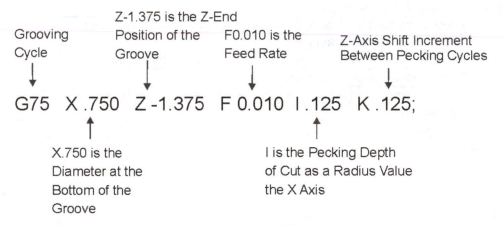

Grooving Cycle

Z-1.375 is the Z-End Position of the Groove

F0.010 is the Feed Rate

Z-Axis Shift Increment Between Pecking Cycles

G75 X .750 Z -1.375 F 0.010 I .125 K .125;

X.750 is the Diameter at the Bottom of the Groove

I is the Pecking Depth of Cut as a Radius Value the X Axis

Figure 10-20. EXAMPLE OF A G75 GROOVING CYCLE.

G75 is the grooving cycle call.

X.750 is the diameter of the groove at the bottom of the groove.

Z-1.375 is the end position of the groove.

F0.010 is the feed rate of the grooving tool.

I is the peck depth of cut in the X axis. Note that it is a radius value.

K is the shift amount on the Z axis. This is used to cut a wider groove or when cutting multiple grooves. In this example to tool is .25" wide and the groove is .375" wide so the K value calls for a shift of .125".

O0918(Program name O0918)

N10 G90 G20 G40 (G90 - Absolute mode, G20 - Inch mode, G40 - Cancel tool-nose radius compensation)

N20 G28 U0.0 W0.0 (Return to reference position - machine home in this example. This is done here to make sure the machine is in the correct location before a tool is loaded. Note that it is common practice to use the U and W value here so that the control does not have to be put into incremental mode for this one line.)

N30 G00 T0505 (Tool 5, offset 5)

N40 G96 S200 M03 (Spindle on clockwise at 200 surface feet per minute constant surface speed)

N50 G00 X1.250 Z-1.250 (Rapid to X1.250 Z-1.250)

N60 G75 X.750 Z-1.375 I.125 K.125 F.010 (Grooving cycle)

N70 G28 X2.00 M05 (G28 - Return to the reference point through X2.00, M05 - spindle stop)

N80 T0500 (Cancel tool offset values in tool 5)

N90 M30 (Program end/memory reset)

%

The G75 grooving cycle is very versatile. It can be used to create one groove that is the same width as the grooving tool, multiple grooves, or grooves that are wider than the tool (see Figure 10-21).

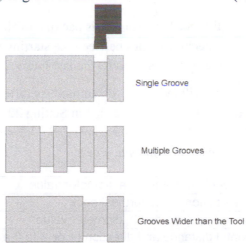

Single Groove

Multiple Grooves

Grooves Wider than the Tool

FIGURE 10–21. Examples of grooves that can be cut with a G75 grooving cycle.

Just like in our first grooving example, the tool must be positioned to the start of the groove prior to calling the grooving cycle. By the use of the cycle's parameters, the controller can be commanded to cut a groove of varying width and depth or multiple grooves. Figure 10-22 shows an example of a part that has three groves that were cut with a G75 grooving cycle. Note that the first groove may not look like it is a groove as it is on the front of the workpiece.

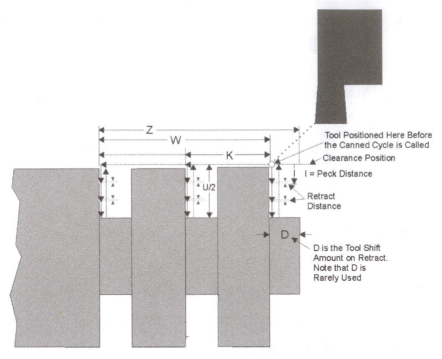

FIGURE 10–22. Example of parameters and their use on multiple grooves.

A G75 canned cycle can be used for grooving an outside diameter. The parameters for a G75 are shown in Figure 10-23. When a Z or W is used in a G75 block and Z is not the current position, then a minimum of two pecking cycles will occur. One is cut at the current Z location and another at the furthest peck location specified by the Z parameter.

The K parameter is the incremental distance between Z axis pecking cycles. The use of the K parameter will perform multiple, evenly spaced, pecking cycles between the starting position and Z. If the distance between S and Z is not evenly divisible by the value in K, the last interval along Z will be less than K.

When the I parameter is used, the peck will feed the peck amount specified in I and then retract in the opposite direction of the feed by the peck distance specified in Setting 22 in a HAAS control. *Note that for any canned cycle you should consult the manual for your particular machine as there are differences in the way canned cycles are programmed in some machines.*

X	X-Axis absolute pecking depth as a diameter value
Z*	Z-Axis absolute location to the furthest peck
U*	X-Axis incremental pecking depth, diameter value
W*	Z-Axis incremental distance and direction (+ or -) to the furthest peck
I*	X-Axis pecking depth increment, radius value
K*	Z-Axis shift increment between pecking cycles
D*	Tool shift amount when returning to the clearance plane (note: rarely used)
F	Feed rate
*	Optional

FIGURE 10–23. G75 grooving canned cycle parameters.

Take a look at the D parameter in Figure 10-22. The D parameter is the tool shift amount when returning to the clearance plane. The D code can be used in grooving and turning to provide a tool clearance shift, in the Z axis, before returning in the X axis to the clearance point. You would not want to use the D command if both sides to the groove exist during the shift, because the groove tool would break. The D parameter is rarely used.

Thread-Cutting Cycle (G76)

The G76 thread-cutting cycle can cut multi-pass threads with one block of information. By using several letter address parameters, the control will automatically calculate the correct number of cut passes, depth of cut for each pass, and the starting point for each pass.

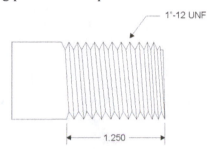

FIGURE 10–24. Example part for a G76 thread-cutting cycle.

Figure 10-24 shows a part with a 1"-12 thread that would be appropriate for a G76 thread cycle.

Figure 10-25 shows an example of a thread cutting canned cycle. To use the G76 thread-cutting canned cycle, the following parameters need to be programmed:

X.900 is the minor diameter of the thread. You can find these values in any reference that shows thread specifications.

Z-1.25 is the absolute Z position of the end of the thread (see Figure 10-22).

I0.0 is the radial difference between the thread starting point and the thread ending point. The I is used for cutting tapered threads. For cutting straight threads, a zero should be programmed.

K.076 is the thread height expressed as a radius value (i.e., [major diameter – minor diameter] divided by 2).

D0.012 is the depth of cut for the first pass (in a radius value). Note: Every pass after the first pass will be decreasing in depth. Figure 10-23 shows the cutting passes for a thread. Note that the first cut depth is larger and subsequent passes get smaller. When we specify D (.012 in our example), the control uses that value and the thread height (K) to determine how many cutting passes should be made to cut the thread. So the smaller the first cut value, the more passes there are that will be made. The controller does this automatically.

If you would like to know how many passes there would be based on your D value, you could use the formula shown in Figure 10-26. There are also charts available on which you select the depth of the first pass and the chart will tell you how many passes will result.

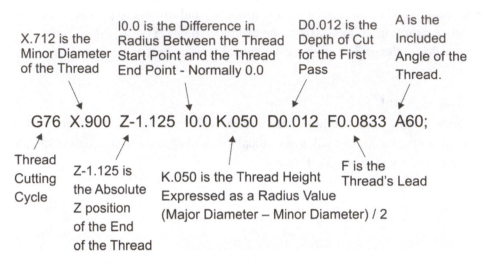

Figure 10-25. Thread cutting cycle to cut the thread shown in Figure 10-24.

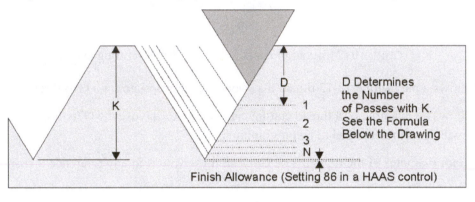

K = Thread Height
N = Number of Passes
D = Depth of First Pass

(K/Square Root of N) = D

Figure 10-26. Diagram of multiple-pass thread cutting.

F is the thread lead (i.e., 1 divided by the number of threads per inch- lead of the thread).

A is the included angle of the thread.

Prior to calling the G76 thread-cutting cycle, the tool must be positioned to the major diameter of the thread plus double the K value or thread height. The tool should also be positioned in front of the thread start position in the Z by a distance of at least double the thread lead. This insures that the proper lead will be cut throughout the length of the thread. The spindle should be running in direct RPM (G97), not constant surface footage control. Following is a program using a G76 thread cycle to cut the thread shown in Figure 10-24.

O0919 (Program name O0919)

N10 G90 G20 G40 (G90 – Absolute mode, G20 - Inch mode, G40 - Cancel tool- nose radius compensation)

N15 G28 U0.0 W0.0 (Return to reference position - machine home in this example)

N20 G00 T0505 (TOOL - 5 OFFSET - 5)

N40 G97 S300 M03 (G97 - Spindle on clockwise at 800 RPM)

N60 G00 Z.200 (Rapid to Z .200 in front of the thread start)

N65 G00 X1.10 (Rapid to X1.10)

N70 G76 X.897 Z-1.25 K.051 D.0120 F.0833 A60 (Threading cycle)

N90 G28 X2.00 M05 (G28 - Return to the reference point through X2.00, M05 - spindle stop)

N100 T0500 (Cancel tool offset values in tool 5)

N110 M30 (Program end/Memory reset)

%

The parameters for a G76 threading cycle are shown in Figure 10-27.

X*	X-Axis absolute thread finish point as a diameter value
Z*	Z-Axis absolute distance, Thread end point location
U*	X-Axis incremental total distance to finish point diameter
W*	Z-Axis incremental thread length finish point
K*	Thread height, radius value
I*	Thread taper amount, radius value
D	First pass cutting depth
P	Thread cutting method P1-P4 (HAAS control - added in version 6.05)
A*	Tool nose angle, no decimal point
F	Feed rate is the thread distance per revolution (Lead of the thread)
*	Optional

FIGURE 10–27. Thread cutting parameters.

G84 Tapping Canned Cycle

The G84 tapping cycle is used to tap a hole on a turning center using a regular tap. Figure 10-28 shows an example of a G84 tapping canned cycle block. This line would rapid to a position .3" in front of the workpiece, control the spindle speed, and then feed the tap to thread a hole 1" deep (see Figure 10-29). The spindle would then automatically reverse direction and the feed reverse to safely feed the tap back out of the work piece to the R plane.

Note: You don't need to start the spindle before the G84 canned cycle. The control turns it on automatically.

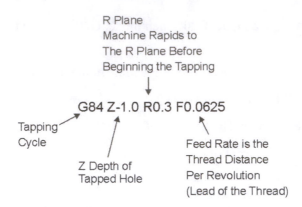

Figure 10–28. Example of a G84 tapping canned cycle call.

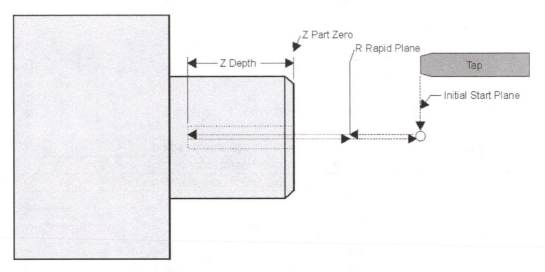

Figure 10–29. Example Of A G84 Tapping Canned Cycle.

The parameters for a G84 tapping cycle are shown in Figure 10-30.

X*	X-Axis absolute rapid location
Z*	Z-Axis depth (Feeding to Z depth from the R plane)
W*	Z-Axis incremental thread length finish point
R	Rapid to R Plane to start feeding
F	Feed rate is the thread distance per revolution (Lead of the thread)
* Optional	

Figure 10-30. Parameters for a G84 Tapping Cycle

O005 (G84 Tapping Program)
N010 G28
N020 T0202 (5-16 Tap) (Tool 2 - Offset 2)
N020 G97 S650 M05 (G84 automatically turns on the spindle)
N030 G00 X0. Z1. M08 (Rapid to Initial Start Point and Turn Coolant On)
N040 G84 Z-1.0 R0.3 F0.0625 (G84 Tapping Cycle)

N050 G80 G00 Z1. M09
N060 G28
N070 M30
%

G85 Boring Canned Cycle

This cycle is also called a bore in-bore out canned cycle. There would have to be a hole in the part before the boring cycle is called. Study Figure 10-31. This boring cycle feeds in at the specified federate until the depth is hit. It then feeds back out to the R plane at the same feed rate.

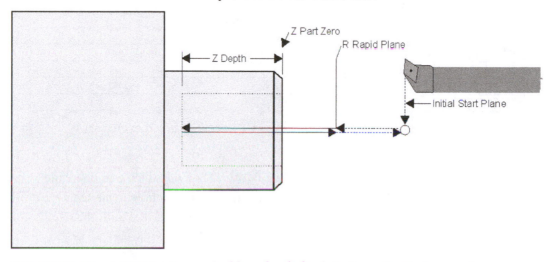

FIGURE 10–31. A G85 boring cycle. Note that it feeds in from the R plane to the programmed depth and then feeds back out to the R plane.

FIGURE 10-32 shows a G85 boring cycle call. Z is the depth of the bored hole. R is the R plane value. In this example it is .2" in front of the part. The tool would rapid to the R plane before beginning the boring feed and then feed out of the hole to the R plane.

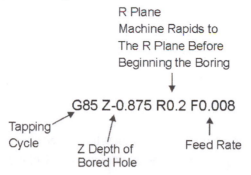

Figure 10-32. A G85 boring canned cycle example.

O008 (G85 Bore in - bore out, program 8)
N010 G28
N020 T0707 (Boring bar) (Tool 7 - Offset 7)
N030 G97 S1200 M03

N040 G00 X0.750 Z1. M08 (Rapid to initial start point)

N050 G85 Z-0.875 R0.2 F0.008 (G85 Bore in - bore out cycle)

N060 G80 G00 Z1. M09

N070 G28

N080 M30

%

Figure 10-33 shows the parameteres for a G85 cycle.

X*	X-Axis absolute rapid location
Z*	Z-Axis depth (Feeding to Z depth from the R plane)
U*	Incremental X axis rapid location
W*	Incremental Z depth (Feeding to Z depth from the R plane)
R	Rapid to R Plane to start feeding
F	Feed rate
*	Optional

FIGURE 10–33. Parameters for a G85 boring canned cycle.

G86 Boring Cycle

A G86 canned cycle is used to bore a hole (see Figure 10-34). There would have to be a hole in the part before the boring cycle is called. The boring cycle is done to machine the hole to an accurate finish size with a good surface finish. This boring cycle feeds in at the specified feed rate until the depth is hit. It then stops feeding and rapids back out to the R plane.

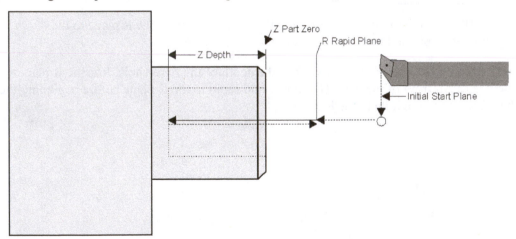

Figure 10-34. A G86 boring canned cycle. Note that it feeds in from the R plane to the programmed depth, stops and then rapids back out to the R plane.

Figure 10-35 shows a G86 canned cycle call. The depth of the bored hole is specified by the Z parameter. The R plane parameter specifies where the machine should rapid to in front of the part. The F specifies the feed rate for the boring cycle.

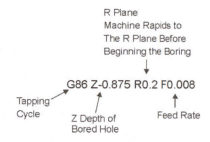

Figure 10–35. A G86 boring canned cycle.

A program that uses a G86 boring cycle follows. Note that it feeds in from the R plane to the programmed depth, stops and then rapids back out to the R plane.

O008 (G86 Bore in-stop- rapid out)

N010 G28

N020 T0707 (Boring bar) (Tool 7 - Offset 7)

N030 G97 S1200 M03

N040 G54 G00 X0.750 Z1. M08 (Rapid to initial start point)

N050 G86 Z-0.875 R0.2 F0.008 (G86 Bore in-stop-rapid out cycle)

N060 G80 G00 Z1. M09

N070 G28

N080M30

%

Figure 10-36 shows the parameters for a G86 boring cycle.

X*	X-Axis absolute rapid location
Z*	Z-Axis depth (Feeding to Z depth from the R plane)
U*	Incremental X axis rapid location
W*	Incremental Z depth (Feeding to Z depth from the R plane)
R	Rapid to R Plane to start feeding
F	Feed rate
*	Optional

Figure 10–36. Parameters for a G86 boring canned cycle.

Now that we have an understanding of canned cycles and how they are used, we need to put our knowledge to work. There are some part drawings after the chapter questions. Use the part drawings and the tool table shown in Figure 10–37 to program these parts. The programs should include canned cycles and tool-nose radius compensation, where appropriate.

Acme Machining Inc.- Tool Table		Part Datum		Part #
Operation	Tool #	Tool Description	RPM	Feed Rate
Rough Turn	T0505	80 Degree Diamond		
Finish Profile	T0303	35 Degree Diamond		
Groove	T1212	.125 Grooving Tool		
Thread	T0707	60 Degree Threading Tool		
Drill 1" Hole	T0909	1" Drill		
Bore Holes	T0808	Boring Bar		

FIGURE 10–37. Tool table.

CHAPTER QUESTIONS

1. True or False? Canned cycles are exactly the same on all CNC machines.

2. True or False? Canned cycles vary on different machine controls so the programmer should check the manual for their particular machine before using a canned cycle.

3. If a part needs to be roughed out of bar stock a G__ would be used.

4. If a part needs to roughed out and finished a G__ and a G__ could be used.

5. Where must the tool be positioned prior to calling a roughing canned cycle?

6. To drill a deep hole in a part a G__ would be used.

7. A G__ could be used to spot drill a hole or to counterbore a hole.

8. Which letter address controls the depth of cut when using canned cycles?

9. What is the letter address command to take .100 of an inch off the diameter of the part per pass when using a roughing cycle?

10. A G__ is used for grooving a part.

11. What letter address controls the pitch or lead of the thread when using a G76 thread-cutting cycle?

12. What must be done prior to the calling of a G76 thread-cutting cycle?

13. Program a roughing and finish canned cycle for the part shown in Figure 10–38. Use canned cycles where appropriate and the tooling shown in Figure 10–37. The rough bar stock is 1.30" in diameter.

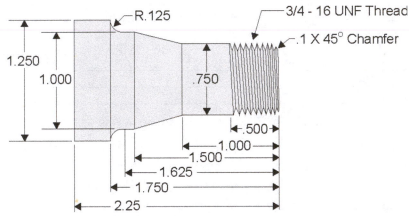

FIGURE 10–38. Use with question 11.

14. Program a threading canned cycle for the part shown in Figure 10–38. Use the threading tool from Figure 10-37.

15. Program a drill canned cycle to drill a 1" hole through the part shown in Figure 10–39. Use the 1" drill from the tool table Figure 10-37.

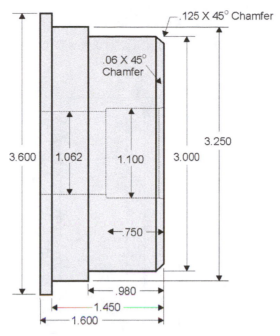

Figure 10–39.

16. Program boring canned cycles to bore the 1.062 and the 1.100 bores in the part shown in Figure 10–39. Use G86 canned cycles. Assume a 1" drill from the tool table (Figure 10-37) was used prior to boring.

17. Program boring canned cycles to rough and finish bore the 1.062 and the 1.100 bores in the part shown in Figure 10-39. Use G71 and G70 canned cycles. Use the boring bar from the tool table shown in Figure 10-37. Assume a 1" drill from the tool table (Figure 10-37) was used prior to boring.

18. Program facing, outside diameter roughing and outside diameter finish canned cycles to machine the part shown in Figure 10-40. Use the tool from the table in Figure 10-37.

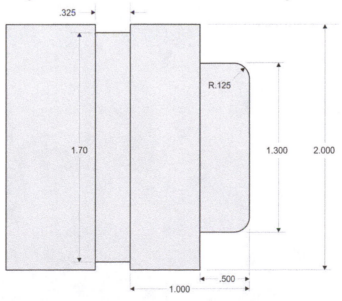

Figure 10-40.

19. Program a grooving canned cycle to machine the grooves on the part in Figure 10-41. Use the tool from the table in Figure 10-37.

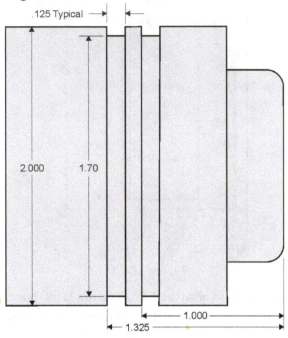

.125 Typical

2.000   1.70

1.000

1.325

Figure 10-41.

Chapter 11

Quality in Manufacturing

INTRODUCTION

This chapter will examine the topic of quality in a typical manufacturing enterprise as well as methods to improve processes. ISO 9001, an international quality standard, will be covered first. Then the basics of statistical quality control will be covered.

OBJECTIVES

Upon completion of this chapter, the reader will be able to:

- Explain What ISO 9001 is and how it impacts a typical machine shop.
- Explain how ISO may affect a machinist.
- Explain calibration.
- Explain the importance of accurate, relevant data.
- Explain terms such as *attribute*, *variable*, and *histogram*.
- Explain the advantages and disadvantages of attribute and variable data.
- Organize data to make it easier to understand.
- Code data
- Explain terms such as "*assignable, chance, special*, and *common causes*."
- Explain the rules of variation.
- Define and calculate averages.
- Define and calculate measures of variation, such as range and standard deviation.
- Explain terms such as *normal distribution* and *bell curve*."
- Predict scrap rates by using statistical means.

ISO 9001

ISO 9001 is an international standard that specifies requirements for a quality management system (QMS). Organizations can use the ISO standard to demonstrate the ability to consistently provide products and/or services that meet customer and regulatory requirements.

The ISO 9001quality standards were first published in 1987 by the International Organization for Standardization (ISO). ISO consists of the national standards bodies of more than 160 countries. The current version of ISO 9001 was released in September 2015. ISO 9001:2015 applies to any organization, regardless of the size or industry.

More than one million enterprises from more than 160 countries have applied the ISO 9001 standard requirements to their Quality Management Systems (QMSs).

The fundamental principle of the standard is that there are some basic things any enterprise must do to be successful and meet their customer's requirements. The standard allows the individual enterprise to create their own system to meet the ISO requirements. This enables companies to develop a Quality Management System (QMS) that meets their own needs and that covers the basic requirements of ISO 9001. To be sure that their systems do meet the requirements of the standard, companies can choose to be certified. To be certified a company hires a third party auditor to come in and make sure their quality management system meets the requirements. If the system does meet requirements and the company is following their system, they can get certified by the third party. The certification is not permanent. They must be audited on a regular basis by the third party auditor to retain their certification.

ISO 9001 is based on the plan-do-check-act methodology. ISO 9001 system is a process-oriented approach to documenting and reviewing the structure, responsibilities, and procedures required to achieve effective quality management in a company.

Changes introduced in the 2015 revision to ISO 9001 were meant to adapt to the changing environments in which organizations operate. Some of the changes in the ISO 9001:2015 revision include the introduction of some new terminology, some structural changes, an emphasis on risk-based thinking, increased leadership requirements, and making it more applicable for service-type enterprises.

The ISO 9001:2015 revision replaced the word *product* with the term *goods and services*. This acknowledges the growth of the service industry, and allows users in service industries to adapt the standard to their unique requirements.

The requirements for documentation have been reduced in the 2015 revision. The terms *documents* and *records* have been replaced with the term *documented information*. The term *continual improvement* has been replaced with *improvement*.

Risk management, change management, and knowledge management are given more emphasis in the ISO 9001:2015 revision. The revision allows organizations greater flexibility and recognizes the need for businesses to integrate their QMS into their overall business strategy.

The ISO 9001:2015 revision is less rigid than previous ISO 9001 versions. The ISO 9001:2015 revision incorporates more business concepts and management terminology. There are fewer documentation requirements and they are much less prescriptive. Companies have more control over what documentation meets their needs.

Benefits of ISO 9001:2015

ISO 9001 helps companies ensure that their customers consistently receive high quality products and services. This can help assure satisfied customers, management, and employees.

ISO 9001:2015 specifies the requirements for an effective quality management system. Companies using ISO 9001 systems have found that the standard helps:

- Organize and improve processes,
- Develop a Quality Management System (QMS),
- Achieve satisfied customers, management, and employees,
- Achieve improvement.

The ISO 9001:2015 standard has 10 clauses.

1. Scope

   What and who the standard covers.

2. Normative references

   This is a blank clause that was used to align the standard's numbering with other management system standards.

3. Terms and definitions

   Definitions of terms referred in the standard.

4. Context of the organization

   Requirements related to determining the purpose and the direction of the organization.

5. Leadership

   Requirements related to the commitment of the top management, quality policy, roles and responsibilities of personnel.

6. Planning

   Requirements related to planning to address risks and opportunities and achieving quality objectives.

7. Support

   Requirements for people, infrastructure, communication, and documented information needed to achieve the purposes of the organization.

8. Operation

   Requirements for identifying customer requirements, design, delivery, and post-delivery support.

9. Performance evaluation

   Requirements related to customer satisfaction, analysis, internal audits and management review.

10. Improvement

    Requirements related to nonconformities, corrective actions and risk-based continual improvement.

A typical ISO 9001 system in a company would have a quality manual (not required in ISO 9001:2015) and procedures that detail how important processes are done and by whom. It should be noted that ISO does not guarantee a good quality system. Some companies have developed quality systems that meet the ISO requirements, but are way too cumbersome and paper intensive. ISO does not require complex, paper intensive systems. The typical worker in a machine shop interacts with the ISO system in several ways. The company's quality system will specify how the operator gets the information to make the parts. This might include blueprints, process sheets, setup sheets, inspection details, and so on. There are probably detailed instructions to the operator on what happens if a piece is scrapped. There might be a form to be filled out, who to report it to, etc.

Calibration

Calibration is one of the main ways an operator interacts with the ISO QMS. To make parts that meet a customer's specifications a company must have accurate gages to inspect them. In addition to the gages being accurate, the operator must know how to use them correctly. For example, if one operator uses a micrometer like a c-clamp they will not get the same readings as another operator that uses a light "feel" when using the same micrometer. Imagine what might happen if a micrometer was to lose its accuracy for a few months. Any operator using that micrometer to inspect parts would have been getting bad readings.

They would have been accepting parts that were out of specification and rejecting parts that were in specification. Out of specification parts would have then been shipped out to the customer. This is very expensive. The customer might inspect and reject the parts. They would then have to wait for replacement parts. The customer might not discover the bad parts and have products fail because of them. The customer might decide they need a new supplier for the parts.

ISO requires that instruments used to inspect parts for quality must be calibrated on regular intervals. Companies can set their own intervals for calibration as long as they prove to be effective. For example, a company might want all micrometers calibrated once each year. Every micrometer would have a sticker on it that would show an identifying number for that micrometer and a date when it is to be calibrated next. The instrument cannot be used if it is beyond the specified calibration date.

The Calibration Log

Each instrument in the calibration system would be entered into a calibration log system. The person responsible for calibration would gather the instruments that needed calibration before their recalibration date. Calibration involves comparing a measuring instrument to standards. A micrometer, for example, might be calibrated by making measurements with very accurate gage blocks and comparing the measurement to the actual size of the gage block. The gage blocks used for calibration would be much more accurate than the micrometer. The calibration gage blocks are usually kept in a secure area and only used for calibrating instruments.

A standard used for calibration is generally at least 10 times more accurate than the instrument being calibrated. The standards are also checked for accuracy on regular intervals.

In many shops the standards would be sent out every few years to companies that would check them against even more accurate standards. This helps assure that every company has standards that are checked against very accurate national standards. When the measuring instrument comes in for its regular calibration it is compared against the standard. If it is accurate it is noted in the log. If it is out of calibration it is also noted in the log. If it is out of calibration it means that the parts it checked between this calibration and the previous calibration may not have been measured correctly.

Many out of specification parts may have been made and sent to the customer. If that is the case, the company must try to assess the situation and remedy anything it can, although that becomes difficult when the parts are gone. The measuring instrument must either be recalibrated or fixed and recalibrated. If it cannot be fixed, it is scrapped and noted in the calibration log. The company should also decide if calibration intervals should be shortened if this has been happening with other similar instruments.

Many companies have operators do a daily or weekly quick check on their measuring instruments against a shop standard between the calibration intervals. This helps prevent the situation where an instrument may have lost its accuracy between calibration intervals.

Fundamentals of Statistics in Manufacturing

Statistics can be a very practical, simple tool when properly applied. The most important fundamental in the effective use of statistics is data. Data must be accurate and relevant. Types and evaluation of data will be examined in this chapter.

Introduction to Statistics

Manufacturing is very competitive. Enterprises have to be competitive in quality, price, and customer service. This means that production processes have to be very efficient and repeatable to make consistent parts in a timely manner. Statistical process control is one way to improve and keep machining processes under control so that they produce accurate, cost effective parts. Machinists should understand the basics of statistics so they can apply them to their work.

How much money is wasted in industry because of the cost of making things wrong? Most experts would say that the average company spends between 10 and 15 percent on the costs associated with doing things wrong. A very successful company might make about 5 percent profit, but many companies are throwing away 2 to 3 times that much due to quality problems!

Cost of Scrap

Imagine we own a small company. We make a widget that costs us 95 cents. We sell the widget for $1 and make 5 cents on every one. What types of expenses go into the 95 cents that it costs to make our widget? Certainly the material cost and the labor cost to make it, and also some of the cost of the machine, tooling, and maintenance. We have to pay the foreman, supervisor, inspectors, sales people, secretaries, president, and so on. We have to pay for the building, utilities, and many other things.

What does it cost if we make a defective widget? Does it cost us our 5 cent profit? Of course! Does it cost us any more than 5 cents? Yes, we lost the 95 cents we had invested in that part, and probably more. We now have to fill out a scrap ticket, our foreman will have to take time to deal with it, and we might have to order more raw materials.

How many parts do we have to make to make up for this loss? We would have to make at least 20 parts: We make 5 cents on each part, so 20 * 5 will make up for the basic loss. We don't make any money on those 20 parts. We just try to recover the loss. While we are making the 20 parts we can't be making anything else, so we lose potential profit there, too. As you can see, the costs of scrap are very high! Statistics and statistical thought can help reduce these costs.

Data

Data is crucial in making manufacturing successful because statistical process control cannot work without accurate data. If we have inaccurate data because of inaccurate measuring equipment (out of calibration) or because people are using it improperly, we cannot make quality parts. There are several important considerations for data collection.

- Make sure that the data is accurate. Consider the gages, methods, and personnel.
- Clarify the purpose of collecting the data. Everyone involved should realize that the purpose is quality improvement. We are not collecting data to make people work harder or get them in trouble.
- Take action based on the data. When we have learned statistical methods, we will only make changes or adjustments to processes based on the data.

The use of data and statistical methods also helps depersonalize problems. Too often, problems become personalized and politicized. No one wants to be blamed for a problem. We would typically prefer to blame it on engineering, or purchasing, or some other department. When we start discussing problems based on data, it depersonalizes a problem. Methods that use data identify problems with processes. The problem should then be discussed in terms of data, not in terms of whose fault it is.

Types of Data

There are two types of data: attribute and variable.

Attribute Data

Attribute is the simplest type of data. The product either has the characteristic (attribute) or it doesn't. If blue is the desired attribute and the product is chairs, we would have two piles of products after we inspected them. One pile would have the desired attribute (blue), and the other pile would have chairs of any other color.

Attribute data can also be go/no-go type data. Go/no-go gages are often used to check hole sizes. If the go end of the gage fits in the hole and the no-go end doesn't, we know the hole is within tolerance.

This type of inspection gives two piles of parts: parts that are within the tolerance and parts that are not (good parts and bad parts). Attribute data does not tell us how good or how bad the parts are. We only know they are good or bad.

Attribute data can provide useful information for decision making.

Attribute data is easy to gather and analyze. Measurements are relatively straightforward for some attributes. A go/no-go gage is almost foolproof; however, some attributes are harder to measure. If the attribute is blue color, it would be easy unless we needed a certain shade. Then it becomes more difficult and subject to error. We can lessen the possibility of error by having a sample to compare against. For example, we might have a sample of the desired shade, one that is too light, and one that is too dark. The inspector can compare the part to the samples if there is any doubt.

Attribute data can also be called *discrete data* because it is either good or bad.

Variable Data

Variable data can be much more valuable. It not only tells us *whether* parts are good or bad, it tells us *how* good or how bad they are.

Variable data is generated using measuring instruments such as micrometers, verniers, and indicators. If we are measuring with these types of instruments, we generate a range of sizes, not just two (good or bad) as with attribute data. We could end up with many piles of part sizes.

This is beneficial because we know how good or bad the parts are. Were they very good or not so good? We can also use this data later to make predictions and decisions about processes.

Errors are possible when we ask people to measure with variable gages. Instruments must be checked against standards on a regular basis to assure they are accurate. This is called *calibration*. We must also make sure all operators are using the same inspection method and that they thoroughly understand the tool they are using for inspection.

Gages must be appropriate for the job. A rule of thumb is that the gage should ideally be able to measure at least 1/10 of the tolerance. For example, if our blueprint specification was 1 inch ±.005, our total tolerance would be .010. Our gage should at least be able to measure to within .001 inch. Variable data can yield more information about a process.

Coding Data

Numbers can become difficult to work with when they have many digits. For example, look at the numbers below. Can you add them in your head and find the average?

> 1.7431 1.7426 1.7437 1.7438 1.7439

Even though there are only five numbers, it is cumbersome. It would be easier if we could work with one-digit numbers. Coding is the way we can do this (see Figure 11–1).

1.005 = 5	1.006 = 6	.999 = -1	.994 = -6
1.009 = 9	1.002 = 2	1.001 = 1	1.000 = 0
.999 = -1	.997 = -3	1.002 = 2	1.001 = 1

FIGURE 11–1. Coded sample part sizes.

Consider the following blueprint specification:

> $1.000 \pm .005$

Let the desired size of 1.000 be equal to zero.

> 1.000 = 0

If we get a part whose size is 1.000, think of it as zero.

We need to measure to the third place to the right of the decimal point (.001). Let that place equal ones. If a part is 1.002, we will call it a 2 because it is two larger than our desired size of 1.000 (zero).

> 1.002 = 2

If the next part were .999, we would call it –1.

> .999 = –1

This is much easier to do than writing down long decimal numbers.

How do you decide which place to assign the value of 1? You simply look at the specification and the number of decimal places. If the specification has three decimal places (5.000, for example), you would make 1 equal one thousandth of an inch. If there were four decimal places, a 1 would equal .0001.

Which numbers would you rather work with: the actual size or a coded value? If you needed to find the average of the numbers, you could just figure the average for the coded values. Because coded values are much easier to work with, most companies use coded values when they use statistical methods.

Graphic Representation of Data

Look at the data in Figure 11–2.

1.023	1.021	1.025	1.021	1.021	1.019	1.023	1.023	1.019	1.022	1.017
1.021	1.020	1.024	1.023	1.012	1.021	1.014	1.022	1.020	1.023	1.023
1.019	1.018	1.022	1.022	1.019	1.021	1.021	1.018	1.019	1.021	1.018

FIGURE 11–2. Some sample part sizes.

What conclusions can you draw? It can be confusing to look at large amounts of data.

It is easier to deal with if we have a picture of it (graphical representation). It would also be easier if it were coded. One method of presenting data graphically is called a *histogram.*

Study the data below. These numbers represent coded values for part sizes that were produced.

1, 4, 1, -1, 1, 3, 2, 3, 3, 2, 2, 2, 4, 1, -1, 1, 1, 0, 0, 1, -1, 0, -1, 0, 0, -2, 0, 0, -1, -3, -2, 1, -4

Now look at Figure 11–3. This histogram makes it easy for us to see patterns in data. You can instantly see that the most common sizes were 0 and 1. We could see that the average part size was between 0 and 1. This histogram is much easier to analyze than the list of data.

The only thing we lose in the histogram is the time value. We do not know if the zeroes were the first parts or the last or if they were distributed throughout the production.

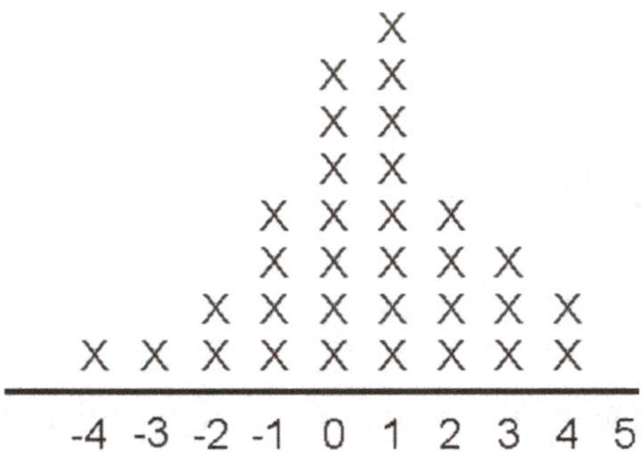

```
                    X
                X   X
                X   X
                X   X
            X   X   X   X
            X   X   X   X   X
        X   X   X   X   X   X   X
    X   X   X   X   X   X   X   X   X
  ─────────────────────────────────────
   -4  -3  -2  -1   0   1   2   3   4   5
```

FIGURE 11–3. Histogram of data from Figure 11–2.

Basics of Variation

Is it true that all things vary and that no two things are exactly alike? Yes. Fingerprints, snowflakes, and so on—no two are exactly alike. All things vary. Look at the earth's climate. It changes day to day, month to month year to year, and milenia to milenia. Variation is normal. The same is true of a stamping die that makes steel washers. If we look at the washers, they all look the same. If we measure them with a steel rule, they all measure the same. But if we use a micrometer, we would find differences in every one. The appropriate measuring instrument will find differences between any two things.

The first rule of variation is straightforward: *No two things are exactly alike.*

Rule number two, *variation can be measured,* is also straightforward. No matter what our product or process, no two will be alike (rule 1), and we will be able to measure and find differences in every part (rule 2).

This assumes that we use an appropriate tool to measure the parts. It also assumes that we have been trained in the correct use of the measuring tool.

The third rule of variation is that *individual outcomes are not predictable.* What would happen if we flipped a coin 10,000 times? We would predict that we would get approximately 5,000 heads and 5,000 tails. We could predict that approximately 50 percent of flips would be heads and 50 percent tails. This would be a very accurate prediction.

What if we flipped the coin 10 times? Our prediction might not be as accurate. We might get seven heads and three tails (not a very good prediction). But if we flipped a larger amount of times, we could predict quite accurately that heads and tails would each occur about 50 percent of the time. Assuming we didn't cheat, 50 percent would be heads and 50 percent tails.

Can we predict that the next flip will be a tail? No! We can never predict an individual outcome.

For example, if we were running a lathe and the last piece was 1.001, can we predict that the next piece will be 1.001? No, because we cannot predict individual outcomes.

Rule number four states that *groups form patterns with definite characteristics*.

Think of a simple process involving a large salt container (see Figure 11–4). The process is to dump 3 ounces of salt from a point 6 inches above the table. The product of the process is the salt piles. If we were very careful, the piles of salt would appear to be the same. The diameter and height would look identical. We could predict the size of the pile if we were to run the process again. In fact, we could make fairly accurate predictions. We could not predict where one grain of salt would fall, however (rule 3). But rule 4 tells us we can make predictions about groups.

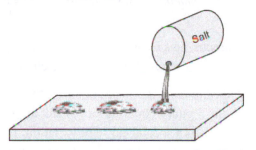

FIGURE 11–4. A simple process.

Would all of the piles be the exact same size? No. Rule 1 tells us no two things are exactly alike.

Why wouldn't each salt pile be exactly the same? There are many reasons.

The height of drop varies. The amount of salt dropped varies. How fast the 3 ounces was poured varies. Air currents in the room change. The humidity in the room changes. The shapes of salt grains vary. The surface of the table is not exactly the same all over.

Some of these reasons would cause large variations, some small. Can you think of any other reasons?

There are many reasons why the piles will not be the same. Statistics tells us, however, that a small number of the reasons cause the majority of the variation. In fact, 20 percent of the reasons will cause 80 percent of the variation. This tells us that if we identify the few key causes, we can dramatically reduce variation. Note also that this applies to all manufacturing processes.

Chance and Assignable Variation

There are two types of causes: chance and assignable.

Assignable causes of variation are those causes that we can identify and fix. For example, the height from which the salt was dropped varied. How could anyone maintain exactly 6 inches of height every time?

Now that we have identified it as an assignable cause, we could make a simple stand to maintain the correct height for the process and eliminate height as a cause of variation.

The amount of salt dropped also varied. Could we expect a person to drop exactly 3 ounces of salt every time? The amount of salt is also an assignable cause. How could we remove this cause of variation?

Statistical methods can be used to identify assignable causes of variation.

Chance causes of variation always exist. Chance causes of variation are those minor reasons which make processes vary. We cannot quantify or even identify all of the chance causes. Room humidity changes may affect the process. Normal temperature fluctuation might affect the process. Normal fluctuations in air currents will affect the process. We really can't separate the effects of these chance causes. Some may even cancel out the effects of others.

You can never eliminate all chance causes of variation. If you are able to identify a cause and its effect, it becomes an assignable cause.

Statistical methods will help us identify assignable causes. Chance causes cannot be separated and evaluated. Chance causes, often called *common causes,* are reasons for variation that cannot be corrected unless the process itself is changed.

For example, if we moved our salt process to a temperature- and humidity-controlled room, we would minimize the effects of humidity and temperature. It takes process change to correct common causes.

Assignable causes are often called *special causes.* These are things that go wrong with a process. For example, the operator starts to pour the 3 ounces but runs out of salt after 1 ounce. The operator refills the container and pours 2 ounces more. This is a special cause of variation. It is something out of the ordinary that occurred in the process. The operator could solve this problem, without a process change, by just making sure the container is full before each pour.

This is the main difference between special and common causes of variation. Special causes are things that are not normal in the process. In other words, something has changed in the process. The operator can often identify these problems and correct them without management action or process change. Common causes of variation are causes that are inherent in the process. The only way to correct the effect of the common causes is to change the process. Only management can change processes; therefore, it is management's responsibility to reduce common causes through process changes.

We now know that variation is normal and that all things vary. We also know that we cannot make predictions about individual outcomes (next flip of the coin), but we can make predictions about groups of outcomes. (If we flip a coin 1,000 times, we will have approximately 50 percent heads and 50 percent tails). We need some ways to describe the characteristics of a group of parts.

- No two things are exactly alike
- Variation can be measured
- Individual outcomes are not predictable
- Groups form patterns with definite characteristics

Average (Mean)

One thing that is helpful in understanding a group of data is what the average or mean is. *Mean* is just another term for average.

For example, if we measured the heights of all persons in a class of 30 students, we could determine the average height. We would add all of the heights up and divide by 30 (the number of students in the class). Statisticians prefer to use the term *mean*. They also have standard notation for data. One person's height, for example, would be called an "x." X stands for one data value. The notation for the average of the data values is $\overline{X}$ (pronounced *x-bar*). Whenever you see a letter with the bar symbol above the letter, it means average or mean. A letter with two bars above it means average of the averages.

Example: heights of five people in inches

x = 60, x = 70, x = 75, x = 80, x = 70   Total = 355/5 = 71 inches

This means that the average of the individual x's was 71 inches. Remember: average and mean are the same.

Measures of Variation

All things vary. It would be desirable to have terms that we could use to describe how much a process varies.

Range

One term used to describe variation is *range*. Range is usually symbolized with the letter "R." The range is simply the largest value in the sample minus the lowest value in the data.

Example 1: A person shopping for a new car looked at five cars. Their prices were $8,000; $15,000; $9,500; $8,500; $12,000. The range of values would equal $15,000 minus $8,000. The range of car prices that this person looked at was $7,000.

Example 2: A food manufacturer measured five consecutive cans of a product and found the weights to be 7.1, 6.9, 7.2, 7.0, and 7.2 ounces. The range would be the highest value (7.2) minus the lowest value (6.9), or .3 ounces.

Range is a very useful, simple value and is a very useful look at variation. It gives us a good quick look at how much a process or sample varies. If we say a process range is .005, we have a good idea of how much the process varies.

Standard Deviation

The second term we can use to describe variation is *standard deviation*. Standard deviation can be thought of as the average variation of a single piece. If we had a sample of pieces and the mean was 10 and the standard deviation was 1, we could say that the average piece varies about 1 from the mean of 10. A statistician would define standard deviation as being the square root of the average square deviation of each variate from the arithmetic mean. (Forget this definition, by the way.) It is more useful to think of standard deviation as the *average variation of a single piece.*

There is an elaborate formula to calculate standard deviation. The formula is not as complex as the definition, but it is tedious and time consuming, and mistakes are easy to make.

Fortunately, most calculators will calculate standard deviation. All you have to do is input the values, and the calculator outputs the mean (average) and standard deviation.

Sample data:

  5, 7, 9, 6, 4, 2, 1, 7, 5, 4

These data values were entered into a calculator, and the standard deviation for the sample was found to be approximately 2.4. The calculated mean was 5.

Find a calculator that figures standard deviation and the manual for it and learn to calculate the standard deviation and mean for the sample.

Calculators can calculate standard deviation for a population or for a sample. Standard deviation for a sample is represented as *S, SD* or *N* – 1. Standard deviation for a population is represented by S, SD or *N*. You should use the standard deviation for a sample, although they would both be very close.

Standard deviation is a very useful term. It can help us visualize processes.

Normal Distribution

If we chose 25 adult men at random and measured their height, some would be very tall and some very short, but most would be about average size. If we plotted a histogram of their heights, it would look something like the one shown in Figure 11–5.

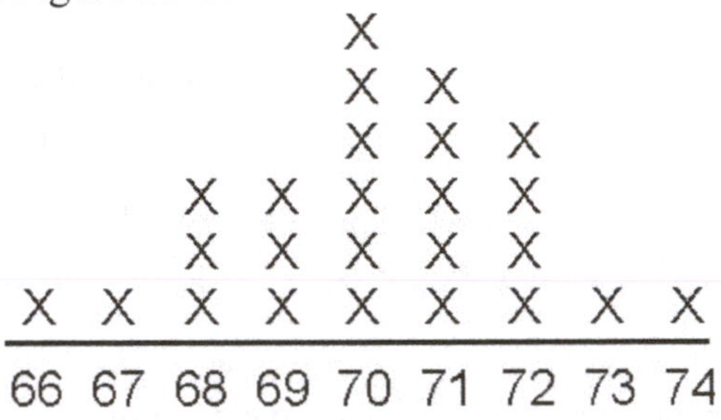

```
            X
            X  X
            X  X  X
      X  X  X  X  X
      X  X  X  X  X
   X  X  X  X  X  X  X  X  X
   ─────────────────────────
   66 67 68 69 70 71 72 73 74
```

FIGURE 11–5. A histogram of the heights of 25 men taken at random. The sizes are in inches.

This distinctive shape is called a *normal distribution* or *bell curve*. A normal distribution looks like a bell.

Most processes produce parts that would be normally distributed. If we measured a dimension on 25 parts from a lathe and then drew a histogram, we would expect a "bell" shape or normal distribution. A few parts would be large, a few small, but most of them would be average.

It should be remembered that we cannot make things exactly alike. There is always variation in processes. The larger the number of parts that we check (or heights of people), the more our histogram would look like a bell shape. More data means more information.

There are some very useful things about normal distributions (bell curves). Consider an example. Assume you measured the height of 25 adult men at random. Their mean height was 5 feet 10 inches (see Figure 11–6). The standard deviation of their heights was also calculated and found to be 2 inches. The bell is normally drawn as being 6 standard deviations wide. In this case, the standard deviation was 2 inches, so the bell would be drawn with a width of 6 * 2 or 12. The bell is then broken into six areas.

The two areas closest to the middle of the bell each contain 34 percent of all the people's heights. The next two each contain 14 percent of the heights. The last two each contain approximately 2 percent. This is very useful because the same relationship exists for any normal process. The percentages will always be the same.

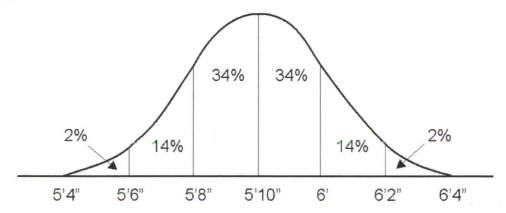

FIGURE 11–6. Bell curve for the heights of 25 randomly chosen men. The mean height for this sample was 5 feet 10 inches tall. The standard deviation was calculated and found to be 2 inches. This means that, based on this sample, 34 percent of all men would be between 5 feet 10 inches and 6 feet tall. We can make all kinds of predictions about men's heights. We could say that approximately 14 percent of all men would be between 5 feet 6 inches and 5 feet 8 inches tall. If our sample was truly random, we could make very good predictions with the data.

Bell curves are set up so that if we know the standard deviation, we can make predictions from the data. Thirty-four percent of all adult males' heights should be between the mean and plus one standard deviation. Thirty-four percent of all adult males' heights should fall between the mean and minus 1 standard deviation. Fourteen percent should fall between plus 1 standard deviation and plus 2 deviations. For this example, we know that the standard deviation equals 2 inches, so we could predict that 68 percent of all adult males will be between 5 feet 8 inches and 6 feet tall.

Statisticians have found that 99.7 percent of all part sizes, heights, or whatever we are measuring will be between ±3 standard deviations.

Consider another example. A lathe is producing pins. The outside diameter was measured on 40 pins from the process as they were being run. The standard deviation was measured and found to be .001. The mean (average) was found to be .300.

Consider the bell curve in Figure 11–7. We could now apply these actual part sizes to the bell curve. The mean is .300. This lets us predict that (assuming we don't change the process) 48 percent of all parts should be between .298 and .300.

What percent would be between .298 and .302? (96 percent); .302 and .306? (approximately 2 percent); .297 and .298? (approximately 2 percent).

This is very valuable information. We can make very accurate predictions about how this process will run in the future.

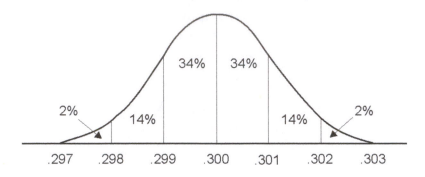

FIGURE 11–7. A bell curve for a process producing parts of sizes .297 to .303 inches. Note that 34 percent are between .299 and .300 inches. Another 34 percent are between .300 and .301. What percentage of parts are between .302 and .303? (2 percent).

Consider the example of people's heights again. Could we make predictions about the heights of all people in Wisconsin from our data? No, because our data was based on adult males only. We could make pretty good predictions about the adult male population of Wisconsin.

The other point to remember is that the more data we have, the better our prediction can be. If we sampled five people's heights, our prediction would not be very good. The larger the sample size, the better predictions we can make.

The term *population* means that we are able to measure all of the items of interest, the entire population. This is a very rare case.

If we produced 1,000 parts on a lathe and measured them all, it would still be a sample if we are going to run more parts. So, we will normally be working with samples, not populations.

We can make very accurate predictions from relatively small samples, and it is more cost effective to use samples. Time is money, and inspection takes time. Statistical methods will help us make predictions that are almost 100 percent accurate.

CHAPTER QUESTIONS

1. What is the ISO 9001:2015 standard?
2. True or false? An ISO 9001 quality management system is the same in every company?
3. True or false? ISO 9001 is an American quality management standard.
4. True or false? A company must be audited on a regular basis by a third party auditor to retain their certification.
5. True or false? Calibration is done to make sure all measuring instruments are accurate in an enterprise.
6. True or false? Out of calibration measuring instruments must be scrapped in an ISO 9001 certified company.
7. True or false? An operator cannot use a measuring instrument that is past its calibration date.
8. True or false? ISO 9001 systems are very paper intensive because everything must be documented.
9. True or false? All measuring instruments that are used to inspect for quality must be included in the calibration log.
10. True or false? Each operator must maintain a calibration log for their measuring equipment.
11. What is attribute data?
12. What is variable data?
13. What are some rules concerning the collection and use of data?
14. What is coding and why is it used?
15. What is a histogram and why is it used?
16. Code the following data: Blueprint specification = 1.126. (Hint: 1.126 = 0.)

Blueprint Specification = 1.126				
1.124 =	1.128 =	1.123 =	1.127 =	1.121 =
1.119 =	1.127 =	1.125 =	1.119 =	1.127 =
1.129 =	1.126 =	1.118 =	1.121 =	1.116 =

17. Code the following:

Blueprint Specification = 1.2755			
1.2752 =	1.2749 =	1.2752 =	1.2754 =
1.2759 =	1.2750 =	1.2761 =	1.2752 =
1.2748 =	1.2756 =	1.2752 =	1.2756 =
1.2753 =	1.2755 =	1.2747 =	1.2749 =

19. Code the following:

Blueprint Specification = 2.105			
2.109 =	2.103 =	2.108 =	2.113 =
2.102 =	2.101 =	2.096 =	2.104 =
2.101 =	2.100 =	2.100 =	2.111 =
2.098 =	2.100 =	2.099 =	2.109 =

20. Code the following:

Blueprint Specification = 5.00				
5.03 =	5.06 =	5.02 =	5.02 =	5.07 =
5.02 =	5.05 =	5.01 =	5.04 =	5.03 =
4.98 =	4.97 =	5.00 =	5.01 =	4.93 =

21. Using the coded data from question 19, construct a histogram.

22. Using the data from question 20, construct a histogram.

23. Define and explain variation.

24. How does an assignable cause of variation differ from a chance cause of variation?

25. Briefly define the following key words.
    a. mean
    b. histogram
    c. normal distribution
    d. bell curve
    e. range

26. Consider the following data. Blueprint specification = 2.250.

1	2	3	4	5	6
2.252	2.249	2.248	2.252	2.246	2.250
2.252	2.249	2.254	2.253	2.248	2.249
2.253	2.243	2.252	2.251	2.247	2.252
2.250	2.248	2.251	2.250	2.246	2.248
2.252	2.249	2.247	2.250	2.249	2.251

a. Code the data. (Hint: 1 should equal 2.251.)

b. Find the sample standard deviation.

27. Blueprint specification = 2.0000.

Subgroup 1	Subgroup 2	Subgroup 3
2.0005	2.0005	2.0002
1.9997	2.0004	2.0004
2.0009	2.0001	2.0000
2.0006	2.0004	1.9995
1.9999	2.0002	2.0001

a. Code the data. (Hint: 2.0005 = 5.)
b. Find the mean.
c. Find the sample standard deviation.

28. Draw a bell curve and label with standard deviations and mean. Label the percentages for each deviation. Note: You don't have data concerning the actual mean and standard deviation. Draw a generic bell curve.

29. Consider the following coded data.

    5, 0, 0, 1, 1, 4, 2, –1, –1, –1, 2, –1, 0, 0, 0, –1, –2, 1, 2, 0, –2, –1, –1, –3, –2

    Use a calculator.
    a. Calculate the mean.
    b. Calculate the sample standard deviation.

30. Consider the following data.

    9, 0, 1, –1, 2, 8, –1, 0, 0, 3, 7, 4, 2, –3, 8, 2, –4, 3, –5, 5, 3, –2, –2, 0, 3

    a. Calculate the mean.
    b. Calculate the sample standard deviation.

31. Consider the following data.

    –3, 2, –1, –2, –4, –1, –2, 0, –2, 3, 0, –1, –1, 4, 2, 2, 0, 0, 1, 4

    a. Calculate the mean.
    b. Calculate the sample standard deviation.

32. A machining process is studied and the mean and standard deviation were calculated: $x = 5$, $s = 2$ (coded data).

    a. Draw a bell curve.
    b. Draw lines where the 6 standard deviations would be.
    c. Label them with actual sizes from this process. (Hint: 99.7 percent of all parts should lie between –1 and +11.)
    d. Label the percentages.

33. Holes are measured after a drill press operation. The mean was found to be 1 (coded data) and the standard deviation was found to be 3 (coded data).

    a. Draw a bell curve for this data.
    b. If 1,000 parts are run, how many will be between –5 and –2?
    c. How many will be between –2 and +4?
    d. If our tolerance was –2 to +7, how many scrap parts would we have?

34. Consider the following coded data.

    3, 3, 3, 3, 5, 4, 2, 4, 2, 1, 5, 4, 3, 2, 4, 6, 1, 4, 3, 2

    a. Construct a histogram.
    b. Does it look like a normal distribution?

35. Consider the following coded data.

    1, 4, 2, 5, 2, 5, 1, 1, 4, 1, 3, 5, 5, 1, 5, 1, 5, 3, 5, 1, 5, 4, 2, 1, 0

    a. Construct a histogram.
    b. Does it look like a normal distribution?
        Note: This one does not look like a normal distribution. This means that something was wrong with our process. Data from a normal process should be normally distributed.

Chapter 12

STATISTICAL PROCESS CONTROL

Control charts are very valuable. If properly used, they can very accurately predict a process's performance. Charts can show when a process has changed, when it should be adjusted, when tools should be changed, when maintenance should be done, and charts can even help find out what has gone wrong with a process.

OBJECTIVES

Upon completion of this chapter, the reader will be able to:

- Explain at least five benefits of charting.

- Explain how charts can improve processes.

- Calculate the capability for a process.

- Construct and properly use $\bar{X}R$ charts.

Process Capability

The real reason to collect data about processes is to use the data to improve our processes. Capability gives us concrete data on how good or bad our processes or machines are.

Consider a simple example. The most common operation on a lathe would be turning an outside diameter to a specific size.

We could perform a study on a lathe to establish the lathe's capability to turn diameters. The term *capability* refers to how close a tolerance a machine can hold. If a lathe can turn diameters to within ±.001, we could say its capability for turning diameters is .002.

To actually conduct the capability study, we would first make sure there is nothing obviously wrong with the machine, tooling, or material. Next, we would run some pieces. Note: We cannot make changes or even adjustments to our process during the study. We want to see how closely the machine can hold sizes, so we must leave the machine alone, make no changes, and just run the parts. The machine is being studied, not your ability to adjust it. It would ruin the results if the machine was adjusted or changes were made while it was running.

For this example, assume 25 parts were run and the outside diameter was measured.

The blueprint mean was 1.125.

The actual sizes of the parts and their coded values are shown in Figure 12–1 (1.125 = 0).

The standard deviation (sample standard deviation) was calculated and found to be 1.67. You should remember (from Chapter 11) that 6 standard deviations are equal to 99.7 percent. If we multiply 6 * 1.67, we get 10.02. (Remember, this is a coded value.)

1st 5	2nd 5	3rd 5	4th 5	5th 5
1.123 = -2	1.124 = -1	1.123 = -2	1.122 = -3	1.127 = 2
1.128 = 3	1.125 = 0	1.125 = 0	1.124 = -1	1.126 = 1
1.127 = 2	1.128 = 3	1.126 = 1	1.125 = 0	1.127 = 2
1.126 = 1	1.126 = 1	1.127 = 2	1.126 = 1	1.125 = 0
1.127 = 2	1.125 = 0	1.127 = 2	1.127 = 2	1.124 = -1

FIGURE 12–1. Diameters and the coded values for the 25 diameters that were cut to study this machine.

This means that if nothing changes in this process, we could run 99.7 percent of all pieces within a range of approximately 10 (coded value). In this case, 1 coded equals .001. We could run 99.7 percent of all pieces on this lathe within approximately .010. (This lathe is obviously not very accurate, but the numbers will be easy to work with for this initial example.)

If the blueprint tolerance is ±.005 (.010 total), we will get almost no scrap (see Figure 12–2).

There is almost no scrap because 6 standard deviations (.010) are equal to our tolerance of .010. (This assumes that we can keep the process exactly on the mean.)

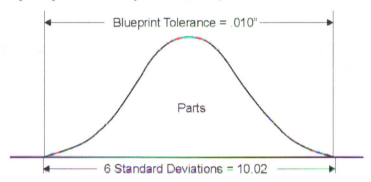

FIGURE 12–2. Bell curve for a job that has a total tolerance of .010 inches and 6 standard deviations (SDs) that are equal to 10.02 inches.

Consider a different job for the same machine. This job has a tolerance of ±.010. This means that the total tolerance is equal to .020 (see Figure 12–3).

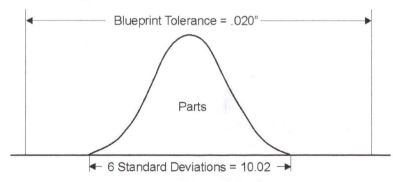

FIGURE 12–3. Bell curve for a job where the process capability is about 1/2 of the blueprint tolerance. This would be a great job, and it would be difficult to produce scrap.

Figure 12-3 makes it clear that this process should produce no scrap. The blueprint tolerance is twice as wide as the process capability.

For an additional example, assume the same process capability (6 SD = .010). This time the blueprint tolerance is equal to ±.003. Study the diagram shown in Figure 12–4. (One standard deviation has been rounded from .00167 to .0017.)

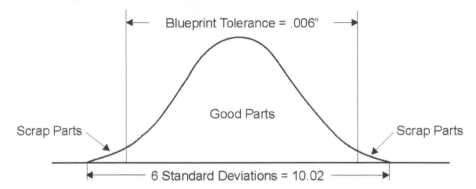

FIGURE 12–4. Bell curve for a process that would produce approximately 4 percent scrap.

In this example, the blueprint tolerance is located at about plus and minus 2 standard deviations. This means that the best this process could do would be about 96 percent good parts (if the process could be kept on the mean). Yelling at the operator will not help reduce the scrap rate. The best that the process can do is 96 percent good parts. Unless the process is changed, it will produce a minimum of 4 percent scrap.

The more parts run, the better our estimate of capability will be. In other words, the more we run, the more confident we could be of our results and the better prediction we could make about running these parts in the future.

Capability is usually not expressed in terms of 6 standard deviations alone.

We would like to be able to compare the process capability to the blueprint specification of the actual job we will be running. This will show how good (or bad) the job is.

One way in which capability is expressed is called *CP* (see Figure 12–5).

$$\text{Capability (CP)} = \frac{\text{Total Blueprint Tolerance}}{\text{6 Standard Deviations}}$$

FIGURE 12–5. The formula for simple capability.

You can see that we are comparing the total print tolerance to the process capability (6 standard deviations) and calling the result CP. Figure 12–6 is used to grade a process for a particular job.

Capability
Above 1.33 – A Excellent
1 to 1.33 – BC
Below 1 - Poor

FIGURE 12–6. Capability chart.

In the first example, the total blueprint tolerance was .010. Six standard deviations was .010. The CP = .010/.010 or 1 (CP = 1). If we look at our capability chart, this job would be classified as a B-C job (not real good, not real bad).

In example two, the blueprint tolerance was .020, and 6 standard deviations was .010. The CP = .020/.010 or 2 (CP = 2). This job is excellent, and it should be easy to run the job without scrap. In fact, it should be almost impossible to run scrap.

In example three, the blueprint tolerance was .006, and 6 standard deviations equaled .010. The CP = .006/.010 = .6. This job is very, very poor. There is no way we can run this process without producing scrap.

CP is one way to express capability. Another way to express capability is easier to use because it is a percentage. The higher the percentage, the worse the job (see Figure 12–7).

$$\text{Capability (CP)\%} = \frac{\text{Total Blueprint Tolerance}}{\text{6 Standard Deviations}} * 100$$

FIGURE 12–7. Formula used to find capability in terms of a percentage. The lower the percentage, the better the capability.

In the first example, 6 standard deviations equaled .010, and total blueprint specification was .010.

.010/.010 * 100 = 100.

The CP% is equal to 100 percent. This means that our process capability is 100 percent of our tolerance.

In example two, 6 standard deviations was .010, blueprint tolerance was .020.

.010/.020 * 100 = 50%

This means that the capability is equal to 50 percent. Six standards deviations is twice as large as the blueprint tolerance, a poor process. Remember: the larger the percentage, the better the job. (The larger the CP%, the smaller the chance of scrap.)

In example 3, the process capability (6 standard deviations) was .010. The total blueprint tolerance is .006.

.006/.010 * 100 = 60%

This is a poor CP%. If our CP% is under 100 percent, we will definitely produce scrap.

Many industries have a goal that CP% should be at least 150%. They will try to improve processes so that they have a CP% of 150% percent or larger. There are many advantages to high CP percentages:

Better parts (parts are closer to the blueprint mean)
Processes are easier to run
Less scrap
Less frequent inspection is necessary
Higher productivity through quality and process improvement

One example of capability comes from the auto industry. One of the automakers had one model of transmission it wanted to study using statistics. The automaker made some of the transmissions and purchased some from another automaker. The transmissions were all made to the exact same specifications and tolerances.

The automaker had noticed that the other manufacturer's transmissions seemed to perform more reliably (less warranty work and customer complaints about noise, and so on). They decided to choose a number of transmissions at random from the other auto-maker and the same number at random from their production. Inspectors were then assigned to completely disassemble the transmissions and inspect them. They checked everything possible, including torque of nuts and bolts and the sizes of all parts.

They found that all of the parts of the other automaker's transmissions met all of the specifications. It was also found that all parts of all their own transmissions met all of the specifications. If all of the transmissions met all of the specifications, why did the other automaker's transmissions perform more quietly and reliably? We would assume that if all parts met specifications, they would perform the same.

This is not true, however. When they analyzed the data, they found that the other automaker had much higher CPs than their own CPs. (In other words, the other automaker's processes had better capabilities.)

Note that they had exactly the same tolerances. The other automaker just had better processes because they used statistical methods and had improved them.

It should be clear that the blueprint mean (center of the blueprint tolerance) should be the ideal size (the size at which the part performs the best). The other automaker made all parts closer to the mean than they did (see Figure 12–8). Thus, their transmissions were quieter and more reliable.

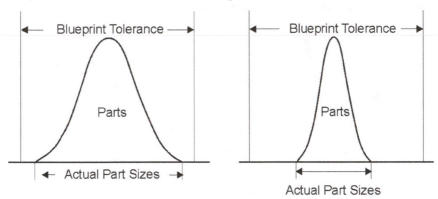

FIGURE 12–8. Two bell curves for two processes making the same part, and the tolerance is the same. The bell curve on the left shows the process bell curve for one company making the part. The bell curve on the right shows the process bell for another company making the same part. Which company makes better parts? Which company has less variation in their parts? Which company's process runs better? Which company's parts would you want in your transmission?

In fact, the other automaker typically had better CPs than many other producers. They accomplished this through process improvement using statistical methods.

Capability, one of the most valuable uses of statistics, tells us much about an enterprise's capability. For example, if we studied all areas of a business and developed capabilities for each machine, we would know what kind of tolerance we could hold on each machine.

It can show us the parts of the business that we need to improve (or even drop). If the lathe department has problems and we cannot hold the tolerances we need to for our customers, this would help us to make the decision to rebuild equipment, buy new equipment, look for work that does not require close tolerances, or decide not to do lathe work anymore. It also can demonstrate an enterprise's strengths.

This can help us bid on work that is more profitable and show us what not to bid on. It can help direct our maintenance efforts and show where to invest in new equipment. It is very useful for the processes where it is difficult or impossible to shift the mean.

Charts can be a very beneficial tool in industry. The charts we will examine next are designed to be accurate 99.7 percent of the time.

Benefits Of Charting

Adjustment Reduction

Charts will show when a machine needs adjustment. People tend to adjust machines too much. In fact, the more conscientious a worker is, the more he or she will adjust. This is because the worker notices the small variations in a process and tries to adjust the machine; however, you cannot eliminate all variation, and by making unnecessary adjustments, you actually make the process variability much worse.

Dr. W. Edwards Deming was one of the most famous quality gurus in history. He is given much of the credit for Japan's amazing manufacturing success. One of Deming's studies involved a paper coating process (see Figure 12–9).

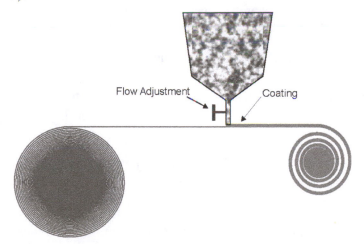

FIGURE 12–9. Simplified paper coating process. The coating thickness is very important: The more consistent the thickness, the better the product.

The coating thickness on the paper was very important. Deming was asked to study the process because it was impossible to keep the thickness consistent.

The worker's job was to measure the thickness and make adjustments when necessary. If the coating was too thick, he closed the valve more; if the coating was too thin, he opened it more. The worker was very diligent.

Deming insisted that the process be run with no adjustments to see how good or bad the process was. Everyone believed that the process would run terribly without adjustment. But Deming persisted, and it was run without adjustment. The process ran very well. There was much less variation in thickness. A chart was put on the machine that showed when adjustment was necessary, and the job became a favorite instead of a problem.

No one knew that variation was normal. The thickness has to vary somewhat. If a machine is adjusted every time a part varies, the process produces much more variation. Charts can make the adjustment decision correctly 99.7 percent of the time.

Process Monitoring

Once a process has been studied and a chart has been constructed, we know how the process should run. The chart will immediately show us when something changes in the process. The chart is 99.7 percent accurate in these tasks.

When we chart, we also are developing historical data on the processes or machines. If we looked at charts over a machine's life, the chart would show the deterioration of the machine or process. If we project this information forward, we could predict when the machine or process will need to be rebuilt or replaced.

Capability

Charts show a business how good or bad their processes and machines are. This seems strange, but many companies do not know how capable their machines are. Charts help decide which work to run on which machines, and they will also identify which machines need rework or replacement.

Charts will also show a company its strengths and weaknesses. For example, a company might find out it is very good at lathe work, but poor on mill work. It can then either improve its mill work by changing processes, maintaining machines, or replacing machines, or it can devote its capital (money) to producing lathe work and letting some other company do the mill work. Whatever they choose, they will be better off than they presently are.

When a company knows how capable its machines are, it can more accurately bid on jobs. The company will also know, before they get the job, how well it will run.

Process Improvement

Charts can become the basis for process improvement. Process improvement means less part variation, and less part variation means higher quality at lower cost. Continual process improvement will yield continually higher quality at continually lower cost.

Imagine the following process: coin flips (heads or tails). A coin is repeatedly flipped. If it comes up heads, an x is plotted above the centerline; tails, an x is plotted below the centerline.

You would expect that approximately half of the flips would be heads and half tails.

The odds of flipping a heads or tails would be 1/2. What are the odds of flipping two heads in a row? The odds would be 1/2 * 1/2, or 1/4. In other words, there is only one chance in four that two consecutive flips would be heads (or tails). The odds for three in a row would be 1/2 * 1/2 * 1/2, or 1/8.

This should help you understand that we would expect half of our product to be above the centerline and half below. We would also expect that we would not get too many in a row on one side of the centerline. The odds of getting seven in a row above (or below) the centerline are 1/128.

This means that if we are plotting sizes of parts on a chart, we would expect that the sizes would occur randomly above and below the centerline.

There is a very small chance that seven in a row would be on one side of the centerline. A process can be adjusted if seven in a row fall on one side of a centerline. Assume seven part sizes in a row fall below center. The odds are so low that this could happen that we could assume that something has changed in the process. This is the second rule: if seven fall on one side of the centerline, the process has changed and an adjustment or change is necessary. If the average of the seven sizes was calculated, it would give the exact adjustment needed. The chart not only tells when to adjust, but also how much to adjust.

In other words, if seven in a row are above or below center, it means that the mean (average) has shifted up or down.

The only other rule is that a process has changed if seven in a row increase or decrease. This is called a *trend*. If each of the seven in a row gets larger (or smaller), this trend means that the process has changed.

These rules are all based on making the correct decision 99.7 percent of the time.

Rule 1: Do not adjust unless a point falls outside of the limits.

Rule 2: Do not adjust unless seven in a row are on one side of the centerline. (If seven are on one side, the mean has shifted.)

Rule 3: Do not adjust the process unless seven in a row trend up or down. Each one must be larger (or smaller) than the last for the trend. This kind of trend will usually indicate something more than an adjustment is needed. Something has changed in the process.

If you follow these rules, you will make the right decision the vast majority of the time.

Charting Processes

The $\overline{X}R$ chart is one of the most widely used charts in industry. The chart is designed to be accurate 99.7 percent of the time.

$\overline{X}R$ means average and range chart. Samples are taken of consecutive parts (usually five) and the average of the sample is plotted on the chart. The range (largest size minus the smallest size) of the sample is also plotted on the chart.

In effect, a chart is really two charts: an average chart with control limits and a range chart with limits.

$\overline{X}R$ Chart Construction

The first step in construction is to gather data from a process. There are often several sizes checked on each piece. One of these must be chosen for a chart. One should try to choose the dimension (or characteristic) that seems most critical. If our process was a lathe part, we might be turning four diameters. Any of the diameters would probably be appropriate to chart. If one diameter were more important, it would be chosen. But if the operations are all very similar, one will be a good indicator of the others.

Once a particular part characteristic has been chosen, data is gathered. It is very important that the process is running well. If we know there is something wrong with the process, it should be fixed before the study is done. The process should be running as well as we think it can (good operator, tooling, etc.).

No adjustments should be made during the study. Because we are trying to study the process, it is imperative that we not change the process through adjustments during the study. We need accurate data on how good the process is, not the operator. The more data that is gathered, the better the results will be. Twenty-five parts will give fairly good results. The data must be consecutive parts.

The data must be accurate. Gages should be appropriate, and the operator should be proficient in using the gage. The operator should also understand why the data is being gathered and the importance of accurate data. If this is not done, there is a tendency for a person to be wary and fudge the data. Remember, we are studying the process, not the person.

Rules:
1. The process should be running optimally
2. Choose an appropriate part characteristic
3. Do not change the process in any way during the study
4. Make sure the operator understands the purpose
5. Use appropriate, well-understood gages
6. Gather the data. (It is a good idea to code the data because it is easier to use and understand. It is also more difficult to make a mistake.)

See Figure 12–10 for data from a process (coded). As you can see, the data was gathered in groups of five.

Subgroup 1	Subgroup 2	Subgroup 3	Subgroup 4	Subgroup 5
0	2	0	2	2
1	0	-1	1	2
1	-3	-1	-2	1
-1	0	-1	1	3
-2	-1	4	0	-2

FIGURE 12–10. Part sizes of the 25 parts that were run. Subgroup 1 contains the first five parts in order, subgroup 2 the next five in order, and so on.

The charts we are constructing work only for normal processes. To see if our process is normal, the data must be put into a histogram (see Figure 12–11). A histogram is a simple, quick look that will indicate whether our process is, or is not, normal.

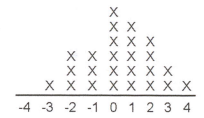

```
              X
              X  X
              X  X  X
           X  X  X  X
           X  X  X  X  X
        X  X  X  X  X  X  X
       _____
       -4 -3 -2 -1  0  1  2  3  4
```

FIGURE 12–11. Data from Figure 12–10 in a histogram. Each X represents one part. For example, in this histogram there were six parts that had a coded size of 0. The histogram also gives us a rough estimate of what the average would be. It also shows the range of sizes produced. In this case, the histogram resembles a bell shape, so the process can be considered normal.

The data in Figure 12–11 looks normally distributed (somewhat like a bell curve). With more data, it would probably look more like a bell. We will assume it is normally distributed. Processes should be normally distributed. If a histogram shows that a process is not normally distributed, something is wrong with the process. The problem in the process must be found and corrected, and new data must be gathered.

We have decided that the process is normal, so we can continue with chart construction. The next step is to find the average (mean) and the range for each group.

First we must find the total for each subgroup. This is done by adding the numbers in each subgroup and writing them in the total column (see Figure 12–12). Study the rest of the subgroup totals.

	Subgroup	Subgroup	Subgroup	Subgroup	Subgroup
	0	2	0	2	2
	1	0	-1	1	2
	1	-3	-1	-2	1
	-1	0	-1	1	3
	-2	-1	4	0	-2
Total	-1	-2	1	2	6
Average					
Range					

FIGURE 12–12. The first step in finding the average for each subgroup. Add the five sizes in each subgroup and write the result in the total box.

Next we must find the average of each subgroup. This is done by multiplying the total of each column by 2 and writing the result in the average column. Then you must put a decimal point one place from the right (see Figure 12–13). The first column total is 1. Multiply 1 * 2, and the result is 2. Next move the decimal place one place to the left and you get .2.

That is the average for the first subgroup; however, this method of averaging works only when there are five values in the subgroup. Study the rest of the subgroup averages.

Subgroup	Subgroup	Subgroup	Subgroup	Subgroup
0	2	0	2	2
1	0	-1	1	2
1	-3	-1	-2	1
-1	0	-1	1	3
-2	-1	4	0	-2
Total -1	-2	1	2	6
Average -.2	-.4	.2	.4	1.2
Range				

FIGURE 12–13. Averages for each subgroup.

The next step is to calculate the range for each subgroup. Remember that the range is simply the difference between the largest and smallest sizes in each subgroup. Look at the first subgroup in Figure 12–14. The largest value is 1 and the smallest value is –2. The difference between these values is 3. Study the other subgroup ranges.

Subgroup	Subgroup	Subgroup	Subgroup	Subgroup
0	2	0	2	2
1	0	-1	1	2
1	-3	-1	-2	1
-1	0	-1	1	3
-2	-1	4	0	-2
Total -1	-2	1	2	6
Average -.2	-.4	.2	.4	1.2
Range 3	5	5	4	5

FIGURE 12–14. The ranges for each subgroup were calculated by subtracting the smallest size from the largest size. For example, the largest size in the first subgroup is 1 and the smallest is –2. The range is equal to the largest minus the smallest, or 3.

The formulas look complex, but are very simple.

The upper control limit for the averages is $UCLx = \bar{\bar{X}} + A_2\bar{R}$.

The upper control limit for the averages is $LCLx = \bar{\bar{X}} - A_2\bar{R}$.

You should notice that the formulas are the same, except that the UCL uses a plus sign and the LCL uses a minus sign.

The formulas really just add an amount (A_2R) to the process average or subtract an amount from the process average.

You already know that $\bar{X}$ is the process average (average of the averages). For our example, it is .24. You can find this by adding the five subgroup averages and dividing by 5.

$\overline{R}$ is the average range for our process. We can find it by adding the five subgroup ranges and dividing by 5, or by adding the five ranges and multiplying by 2 and moving the decimal place one place to the left. For this example, the average range is 4.4.

If we substitute these into the formula, we have:

$$UCL_{\overline{X}} = .24 + A_2 * 4.4$$
$$LCL_{\overline{X}} = .24 - A_2 * 4.4$$

Next, we need to know what A_2 is. A_2 is a constant from a table (see Figure 12–15).

Number in Subgroup	Value of A2	Value of D4
3	1.02	2.57
4	.73	2.28
5	.58	2.11
6	.48	2.0
7	.42	1.92
8	.37	1.86
9	.34	1.82
10	.31	1.78

FIGURE 12–15. This figure shows how the values for A_2 and D_4 are found in the chart. There are five pieces in our subgroups, so the value of .58 is used.

The number of parts in our subgroup is 5, so the value for A_2 (n) is .58.

We just substitute .58 into the formula.

$$UCL_{\overline{X}} = .24 + .58(4.4) = 2.792 \quad LCL_{\overline{X}} = .24 - .58(4.4) = -2.312$$

Make sure to perform the multiplication first, then calculate the two limits. They should be $UCL_{\overline{X}} = 2.792$ and $LCL_{\overline{X}} = -2.312$.

This completes the averages portion of the $\overline{X}R$ chart.

The formula for the upper control limit for the range (UCL_R) is $UCL_R = D_4\overline{R}$. The value of $D_4\overline{R}$ is found in the chart in Figure 12–15.

There are five pieces in our subgroups, so we will use a value of 2.11 for D_4.

The formula says multiply D_4 by the average range ().

$UCL_R = 2.11 * 4.4$

$UCL_R = 9.284$

A process will always have variation. The upper control limit on the range will tell us when the variation is higher than it should be. (If a range is greater than the upper limit, there is a 99.7 percent chance that something changed in the process.)

Figure 12–16 shows a completed chart for this example.

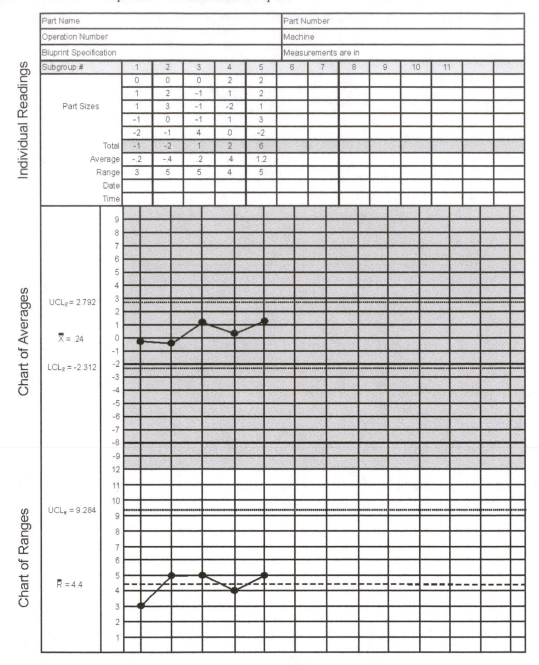

	Part Name					Part Number					
	Operation Number					Machine					
	Bluprint Specification					Measurements are in					

Subgroup #	1	2	3	4	5	6	7	8	9	10	11
Part Sizes	0	0	0	2	2						
	1	2	-1	1	2						
	1	3	-1	-2	1						
	-1	0	-1	1	3						
	-2	-1	4	0	-2						
Total	-1	-2	1	2	6						
Average	-.2	-.4	.2	.4	1.2						
Range	3	5	5	4	5						
Date											
Time											

Individual Readings

Chart of Averages

$UCL_{\bar{x}} = 2.792$

$\bar{\bar{X}} = .24$

$LCL_{\bar{x}} = -2.312$

Chart of Ranges

$UCL_R = 9.284$

$\bar{R} = 4.4$

FIGURE 12–16. Completed chart.

The data was transferred to the chart. Notice that the chart really has three areas. The top area contains information about the individual parts: the part name, part number, operation number, and so on, as well as the actual sizes of the parts that were made.

The actual part sizes were written down in order in the subgroup areas in groups of five. This area is also used to calculate the averages and ranges for each subgroup.

252

The middle portion of the chart is used to plot the average for each subgroup. This area is the chart of averages. Notice that the mean and the upper and lower control limits have dotted lines to mark the limits. The actual values for the limits are written on the left side of the chart.

If a point that we plot is between the limits, the process is acceptable. The chart will always tell us if something has changed in the process. If a point falls outside the limit, there is a 99.7 percent chance that something has changed in the process.

The subgroup averages are plotted by following the vertical line under each subgroup down to the average portion of the chart. The average for subgroup 1 is –.2. Follow the line under subgroup one down and you will see that a dot was drawn on the line just under 0.

The second subgroup's average was –.4. Follow the line down from the second subgroup and you will see that a dot is drawn at –.4. The third subgroup has an average of 1.2. After all of the five averages were plotted, they were connected with lines.

The third area of the chart is the chart of ranges. The range portion of the chart is used to graphically show variation between parts in a subgroup. A small range is desirable and indicates that the variation between parts within the subgroup is small.

Note there is only an upper limit on the range chart. We want all of our subgroup ranges to be under the limit. Remember that the smaller the range the better, because it means that there is less variation between the parts in the subgroup.

The subgroup ranges are plotted by following the vertical line under each subgroup down to the range portion of the chart. The range for subgroup 1 is -3. Follow the line under subgroup 1 down and you will see that a dot was drawn at 3 in the range portion of the chart. The second subgroup's range was 5. Follow the line down from the second subgroup and you will see that a dot is drawn at 5. The third subgroup has a range of 5. After the five ranges were plotted, they were connected with lines.

Analyzing the Chart

Are all of the subgroup averages inside of the upper and lower control limits? Are all of the subgroup ranges under the upper control limit? All of the subgroup averages were inside the limits, as were the subgroup ranges.

Remember, the chart makes correct decisions 99.7 percent of the time. If we use the chart, we will make good decisions 99.7 percent of the time. However, we have not even looked at blueprint tolerance for this job yet.

A chart assures us that a process is running at its best, but even the best may not be good enough for a very tight tolerance. The operator must still watch the individual piece sizes and scrap the bad parts. For example, how many parts would be scrap for this job if our tolerance was ±.003 inches? You should find one scrap part. Note that the operator must scrap parts that are outside of the blueprint tolerance, but the data must be entered in the chart. Also, notice that the points we plotted did not tell us we had scrap.

Examine the chart in Figure 12–16. Are all of the averages inside the limits? All of the ranges?

The chart would now be ready to be used on this job. Note that we only calculate limits once when we first study the job. From this time on, we would just enter the date and plot points as we run parts.

The chart assures us that the process (or machine) is running at its best. It does not assure us there is no scrap. The operator must compare the individual parts to the blueprint tolerance. If the particular job we are running has a very good CP, we can reduce our inspection. The better the job (wide tolerance compared to six or more), the less we need to inspect. We might be able to sample one subgroup an hour or one a day if we found an excellent job.

Operators should be encouraged to write notes about the job as they run it. Even hunches could prove useful. The more notes on charts the better. If we examine a chart months or even years later, notes will improve our recall.

Now that limits have been set for this particular job on this particular machine, they never have to be recalculated again unless we change the machine or process.

More parts were run on another day. They were entered into the chart, and the averages and ranges were calculated and plotted (see Figure 12–17). Note that we did not have to recalculate limits. If we run the same job on the same machine and with the same tooling, we can use the same limits and the job should run the same. This is very beneficial because we are developing some history on the job and machine. If we were to look at the same job over a long period of time, we might be able to see the variability increase as the machine wears. This might help us plan maintenance and machine replacement.

Note that in subgroup 7 there is an ellipse drawn around the –6 part. This means that there was a tool change done, which explains why the size was so far off. Note that the average for this group was all right, but the range was outside the limit. The range shows that the variability for this subgroup was high because of the tool change, which the operator noted on the chart.

The average for subgroup 8 is well within the limits. The range is fine also.

The average for subgroup 9 is above the upper limit. The operator stopped running parts when the average exceeded the limit. The chart is telling us that something changed. The operator studied the machine, the setup, and the tooling and discovered that the tool was chipped. The operator noted this on the chart and initialed the note. Notes are very important on charts. The range for the subgroup was within the limit.

You should begin to see that a chart will instantly tell us if there is a problem. The chart will make the right decision 99.7 percent of the time. The operator should not make any changes or adjustments to the machine unless the chart indicates a change has occurred in the process.

Rule 1: Do not make any changes to a machine unless the average or range is outside of the limits.

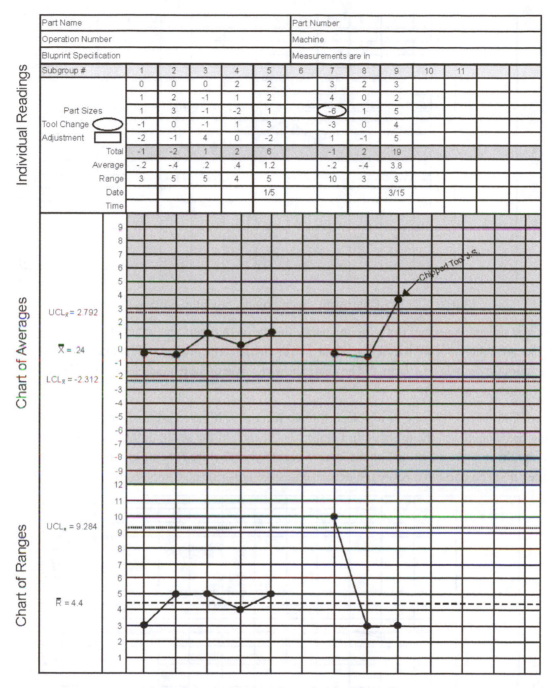

Part Name						Part Number				
Operation Number						Machine				
Bluprint Specification						Measurements are in				
Subgroup #	1	2	3	4	5	6	7	8	9	10	11
	0	0	0	2	2		3	2	3		
	1	2	-1	1	2		4	0	2		
Part Sizes	1	3	-1	-2	1		-6	1	5		
Tool Change	-1	0	-1	1	3		-3	0	4		
Adjustment	-2	-1	4	0	-2		1	-1	5		
Total	-1	-2	1	2	6		-1	2	19		
Average	-.2	-.4	.2	.4	1.2		-.2	-.4	3.8		
Range	3	5	5	4	5		10	3	3		
Date					1/5				3/15		
Time											

FIGURE 12–17. Completed chart.

Trends

The second rule for charts is that something has changed in the machine or process if there are seven points in a row either above or below the average (see Figure 12–18). This partial chart shows the plots of seven points. Note that all seven are above the average. This indicates something has changed in the process. The operator should stop the machine and see if anything obvious is wrong. If nothing is found, an adjustment should be made.

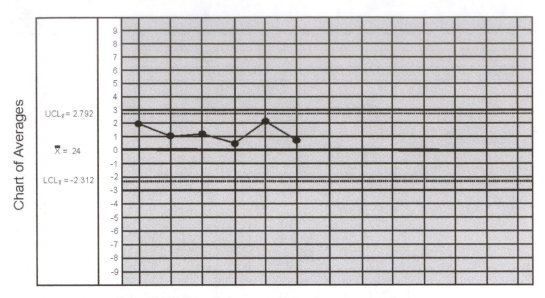

FIGURE 12–18. Chart with seven points above the process average.

If you look at the averages and draw an imaginary line through them, you would see that their average is about 1. Because the parts are averaging about 1 over size, the operator should make an adjustment of 1. This is a major advantage of charting. The chart shows us exactly how much to adjust. (This rule also applies to subgroup ranges.)

The second kind of trend is one in which each point increases. If seven points increase in size in a row, something has changed in the process. In Figure 12–19, each of the seven points increased in size. The operator should stop the machine and discover what is wrong before running any more pieces. The rule also applies if seven parts in a row each decrease in size. This rule also applies to the subgroup ranges.

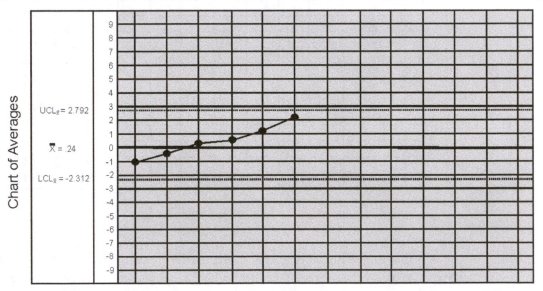

FIGURE 12–19. Chart with a trend of seven subgroup averages in a row increasing in size.

Do not change anything unless:

1. An average or range is outside of the limits

2. Seven averages or ranges in a row are above (or below) the average

3. Seven averages or ranges in a row increase or decrease

Charts can be invaluable if used correctly. A shop that understands and uses statistical methods will be much more successful than a shop that does not.

Note that even if you do not use charts, a fundamental understanding of statistical principles will improve quality and productivity.

CHAPTER QUESTIONS

1. List and explain the three rules for charts that show when a process has changed.
2. Explain at least three benefits of charts. How do they help operators, maintenance, bidding, and so forth?
3. These charts make correct decisions _____ percent of the time.
4. Why is it important that no adjustments be made to the machine or process during the initial study?
5. A machine has a standard deviation of .003. We are considering running a job on the machine that has a tolerance of ±.006.
   a. Calculate the CP.
   b. Will there be scrap? If so, how much? (Hint: Draw a bell curve to help find the answer.)
6. Using the same machine as in question 5, assume the job has a tolerance of ±.010.
   a. Calculate the CP.
   b. Will there be scrap? If so, how much?
7. Standard deviation = .0015; blueprint tolerance = ±.005.
   CP = _____
8. Standard deviation = .001; blueprint tolerance = ±.002.
   CP = _____
9. Based on the data from question 8, will this job run well? If you were asked whether to bid on the job, list at least three alternatives you could give.
10. You are asked to make a recommendation on bidding on a lathe job. The company supplied you with 30 sample pieces to run. The data for the job is shown in Figure 12–20. Code the data and draw a bell curve that compares the capability (6 standard deviations) to the blueprint specifications. Blueprint Specification = 2.250 ± .004.

Subgroup 1	Subgroup 2	Subgroup 3	Subgroup 4	Subgroup 5	Subgroup 6
2.252	2.249	2.248	2.252	2.246	2.250
2.252	2.249	2.254	2.253	2.248	2.249
2.253	2.243	2.252	2.251	2.247	2.252
2.250	2.248	2.251	2.250	2.246	2.248
2.252	2.249	2.247	2.250	2.249	2.251

FIGURE 12–20. Use with question 10.

11. Complete an $\overline{X}R$ chart for the process in question 10 (enter the data, calculate control limits, and plot the data).
    a. Standard deviation
    b. Process mean
    c. Average range
    d. Is the data normally distributed? (Check with a histogram.)
    e. Upper control limit averages
    f. Lower control limit averages
    g. Upper control limit range

Part Name						Part Number						
Operation Number						Machine						
Bluprint Specification						Measurements are in						

Subgroup #	1	2	3	4	5	6	7	8	9	10	11		
Part Sizes													
Total													
Average													
Range													
Date													
Time													

Individual Readings

Chart of Averages

$UCL_{\bar{x}} =$

$\bar{\bar{X}} =$

$LCL_{\bar{x}} =$

Chart of Ranges

$UCL_R =$

$\bar{R} =$

12. The data in Figure 12–21 was taken from 25 consecutive pieces. The process was a turning operation on a lathe. The blueprint specification was $1.250 \pm .003$.

   a. Code the data.
   b. Is the process normal?
   c. Mean.
   d. Average range.
   e. Complete an $\overline{X}R$ chart (enter the data, calculate control limits, and plot the data).
   f. Compute the capability.

Subgroup 1	Subgroup 2	Subgroup 3	Subgroup 4	Subgroup 5
1.253	1.254	1.251	1.249	1.247
1.250	1.249	1.245	1.250	1.251
1.247	1.250	1.251	1.256	1.252
1.251	1.253	1.248	1.252	1.251
1.248	1.249	1.252	1.248	1.245

FIGURE 12–21. Use with question 12.

Appendix A – Typical Machining Center M and G-Codes

Common M Codes		
Note that these codes vary between manufacturers and even between models of CNC machines. Check the manual for your particular control.		
M00	Program stop	Non-Modal
M01	Optional stop	Non-Modal
M02	End of program	Non-Modal
M03	Spindle start clockwise	Modal
M04	Spindle start counterclockwise	Modal
M05	Spindle stop	Modal
M06	Tool change	Non-Modal
M07	Mist coolant on	Modal
M08	Flood coolant on	Modal
M09	Coolant off	Modal
M30	End of program & reset to the top of program	Non-Modal
M40	Spindle low range	Modal
M41	Spindle high range	Modal
M98	Subprogram call	Modal
M99	End subprogram & return to main program	Modal

Common Machining Center G Codes

Note that these codes vary between manufacturers and even between models of CNC machines.

Make sure to check the manual for your particular control.

Code	Description	Modal
G00	Rapid traverse (rapid move)	Modal
Example: G00 X## Y## Z## F## (X,Y,Z = Position)		
G01	Linear positioning at a feed rate	Modal
Example: G01 X## Y## Z## F## (F=Feed Rate)		
G02	Circular interpolation clockwise- IJ example	Modal
Example: G02 X## Y## I## J## F## (XY = End Point, IJ = Start to Center		
G02	Circular interpolation clockwise – radius example	Modal
Example: G00 X## Y## R## F## (XY = End Point, R = Radius, + if Radius <180°, - if Radius > 180°		
G03	Circular interpolation counter-clockwise	Modal
Example: G03 X## Y## I## J## F## (XY = End Point, IJ = Start to Center		
G03	Circular interpolation counter-clockwise	Modal
Example: G00 X## Y## R## F## (XY = End Point, R = Radius, + if Radius <180°, - if Radius > 180°		
G04	Dwell	
Example: G04 P## (P=Time, P20000 = 2 Seconds)		
G09	Exact stop positioning move	
Example: G09 X## Y## Z## F## (Active for single block)		
G10	Zero Offset Shift	
Example: G10 X## Y## Z## (X, Y, Z Shift Distance)		
G15	Turn polar coordinates off, return to Cartesian coordinates	
Example: G15		
G17	XY Plane Selection	Modal
Example: G17		
G18	ZX Plane Selection	Modal
Example: G18		
G19	YZ Plane Selection	Modal
Example: G19		
G20	Inch Mode	Modal
Example: G20		
G21	MM Mode	Modal
Example: G21		
G28	Zero or home return	
Example: G00 G91 G28 X## Y## Z## (Go to Machine Home passing through X,Y,Z Incremental Mode)		

G29	Return from Reference Point	
Example: G00 G90 G29 X## Y## Z## (Go to This XYZ Position, Returning from Home)		
G30	Return to 2nd, 3rd (etc.) Reference Point	
Example: Similar to G28		
G40	Tool diameter compensation cancel	Modal
Example: G40 X## Y##		
G41	Tool diameter compensation-left	Modal
Example: G41 X## Y##		
G42	Tool diameter compensation-right	Modal
Example: G42 X## Y##		
G43	Tool height offset – Spindle Approach from + Side	Modal
Example: G43 H## Z##		
G44	Tool height offset – Spindle Approach from - Side	Modal
Example: G44 H## Z##		
G45	Increase end position by tool offset value	
Example: G45 X## D## (Go to X position plus offset value in D##)		
G46	Decrease end position by tool offset value	
Example: G46 X## D## (Go to X position less the offset value in D##)		
G47	Increase end position by twice the offset value	
Example: G47 X## D## (Go to X position plus twice the offset value in D##)		
G48	Decrease end position by twice the offset value	
Example: G47 X## D## (Go to X position less twice the offset value in D##)		
G49	Tool height offset cancel	Modal
G53	Coordinate system referenced from machine home	
Example: G53 X## Y## Z## (Go to this XYZ position referenced from machine home)		
G54	Workpiece coordinate preset (shift), offset 1	
Example: G54 X## Y## Z## (Go to this XYZ position referenced from work coordinate preset #1)		
G55-G59	Workpiece coordinate preset (shift), offset G55-G59	
Example: G%# X## Y## Z## (Go to this XYZ position referenced from work coordinate preset G55 – G59)		
G61	Exact stop cutting mode	Modal
Example: G61 X## Y## Z## (Decelerated at point XYZ, before next		
G62	Feed compensation on inner corner	
Example: G62 G02 X## Y## I## J##		

G63	Feed override lock out	
Example: G63 X## Y## Z##		
G64	Exact stop mode off	
Example: G64 X## Y## Z## (Tool is not decelerated at point XYZ)		
G73	Rapid Peck Cycle - The G73 cycle is used for deep drilling or milling with chip breaking. The tool retracts to break the chip but does not totally retract the drill from the hole.	Modal
Example: Varies by manufacturer		
G80	Canned cycle cancel	Modal
Example: G80		
G81	Canned drill cycle – feed in, rapid out	Modal
Example: G81 X## Y## Z## R## F##		
G82	Counter bore cycle- feed in, dwell, rapid out	Modal
Example: G82 X## Y## Z## R## F## P####		
G83	Canned peck drill cycle- feed in peck amount, rapid in within .05 of last peck and repeat until depth is reached	Modal
Example: G83 X## Y## Z## R## F## Q##		
G84	Canned tapping cycle- feed in, spindle stop, reverse, feed out	Modal
G84 X## Y## Z## R## F##		
G85-G89	Canned boring cycles, function and inputs vary between manufacturers	Modal
Vary by manufacturer		
G90	Absolute coordinate positioning- Point based from XYZ zero	Modal
Example: G90 G00 X## Y## Z##		
G91	Incremental positioning- Point to point dimensioning	Modal
Example: G91 G00 X## Y## Z##		
G92	Absolute zero preset- The current position is set to the value shown in the line	
Example: G92 X15 Y8 Z-7 (After this command is run the current position of the machine will be X15 Y8 Z-7. (Using G54-G59 is preferred to the G92 method)		
G94	Feed rate is read as inches per minute	Modal
Example: G94		
G98	Canned cycle initial point return- retract the tool to the starting Z point when drilling. Used for high retract clearance when moving between holes. For example if the initial Z height was 1.5" the drill cycle below would rapid to the R plane, drill to the Z depth and return to the starting height of 1.5" when the cycle is finished.	Modal
Example: G98 G81 X## Y## Z-1.0 R.100 F##		

| G99 | Canned cycle R point return, Retract the tool to the R plane when the drilling cycle is finished. The tool will return to the R plane height in the program line. | Modal |

Example: G99 X## Y## Z-1.0 R.100 F##

265

APPENDIX B – TYPICAL TURNING CENTER M AND G-CODES

M00	Program stop
M01	Optional stop
M02	End of program
M03	Spindle start clockwise
M04	Spindle start counterclockwise
M05	Spindle stop
M07	Mist coolant on
M08	Flood coolant on
M09	Coolant off
M30	End of program & reset to the top of program
M98	Subprogram call
M99	End subprogram & return to main program

G00	Rapid traverse (rapid move)
G01	Linear positioning at a feed rate
G02	Circular interpolation clockwise
G03	Circular interpolation counter-clockwise
G04	Dwell
G09	Exact stop
G10	Programmable data input
G20	Input in inches
G21	Input in mm
G22	Stored stroke check function on
G23	Stored stroke function off
G27	Reference position return check
G28	Return to reference position
G32	Thread cutting
G40	Tool diameter compensation cancel
G41	Tool diameter compensation-left
G42	Tool diameter compensation-right
G50	Max. spindle speed clamp
G70	Finish Turning Cycle
G71	Rough Turning cycle
G72	Facing cycle
G73	Pattern repeating cycle
G74	Peck drilling cycle
G75	Grooving cycle

G76	Threading cycle
G80	Canned cycle cancel
G81	Drill caned cycle
G82	Spot drill canned cycle
G83	Peck drilling caned cycle
G84	Tapping canned cycle
G85	Boring canned cycle
G86	Bore and stop canned cycle
G95	Feed per revolution
G96	Constant surface speed control
G97	Constant surface speed control cancel

CHAPTER 1 QUESTIONS

1. Describe the 3 axes that a milling machine typically has.

A milling machine typically has 3 axes of motion. The X axis is the table movement right and left as you face the machine. The Y axis is the table movement toward and away from you. The Z axis is the spindle movement up and down. A move toward the work is a negative Z (–Z) move. A move up in this axis would be a positive Z (+Z) move.

Milling machines use all three axes, as seen in Figure C–1. The X axis usually has the longest travel. On a common vertical milling machine, the X axis moves to the operator's left and right. The Y axis moves toward and away from the operator. The Y axis usually has the shortest travel. The Z axis always denotes movement parallel to the spindle axis, the up and down movement. Toward the work is a negative Z move. It may be helpful to think of the motion in terms of tool position. If the table is moved to the left, the tool is positioned more in the + X direction. If the table is moved toward the front, the tool is positioned more in the –Y direction.

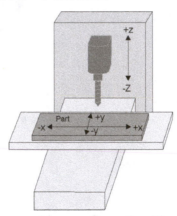

FIGURE C-1. Typical milling machine configuration illustrating the X, Y, and Z axes.

3. Describe ball screws and explain why they are used on CNC machines.

The rotary motion generated by the drive motors is converted to linear motion by recirculating ball screws. The ball leadscrew uses rolling motion rather than the sliding motion of a normal lead screw. The balls, located inside of the ball screw nut, contact the hardened and ground lead screw and recirculate in and out of the thread (see Figure C–2). The contact points of the ball and screw directly oppose one another and virtually eliminate backlash. The contact points are also very small, so very little friction is generated between them.

Other advantages of the ball lead screw over the Acme lead screw are:
*   less wear
*   high speed capability
*   precise position and repeatability
*   longer life

5. Describe the incremental positioning mode.

Incremental programming specifies the movement or distances from the point where you are currently located (see Figure C–3). A move to the right or up from this position is always a positive move (+); a move to the left or down is always a negative move (–). With an incremental move, we are specifying how far and in what direction we want the machine to move.

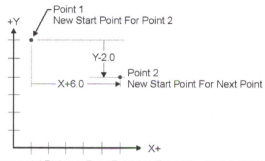

Incremental Distance From Point 1 to Point 2 is   X+6.0, Y-2.0

FIGURE C–3. In incremental coordinate positioning, your present position becomes the program start position. From the start point, point 1 is one position to the right on the X axis (X+1) and six positions up on the Y axis (Y+6). When we move to point 2, point 1 becomes our start point. Point 2 is 5 positions to the right of our present position on the X axis (X+5) and 2 positions down on the Y axis (Y–2). Remember, in incremental programming, moves down or moves to the left are negative moves.

7. Identify the positions marked on the Cartesian coordinate grid below. Use absolute positioning.

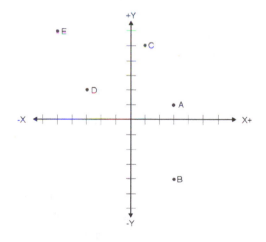

Point	X	Y
A	3	1
B	3	-4
C	1	5
D	-3	2
E	-5	6

Chapter 2 Questions

1. Explain what is meant by the term cutting speed.

   Different materials need to be cut at different speeds. The term used to describe a material's ideal speed for cutting is cutting speed. Cutting speed is the speed at the outside edge of the cutter as it is rotating. This is also known as surface speed. Surface speed, surface footage, and surface area are all directly related. Next imagine different types of materials: plastic; aluminum; brass; mild steel; alloy steel; hardened steel, and so on. Cutting speeds are given in surface feet per minute (SFPM).

3. What is the formula for calculating RPM?

$$\frac{Cutting\ Speed\ (CS) * 4}{Diameter\ of\ Cutter\ (D)} = RPM$$

5. State the two main characteristics of carbide.

   Hard and brittle.

7. What is a coated carbide?

   Carbide is a very hard, durable cutting tool, but it still wears. The wear resistance of cemented carbide can be greatly increased by using coated carbide inserts. Wear-resistant coatings can be applied to the carbide substrate (base material) through the use of plasma coating or vapor deposition. The coating is very thin but very hard. The most common types of coatings include titanium carbide (TiC), titanium nitride (TiN), and aluminum oxide (AlO). Aluminum oxide is a very wear-resistant coating used in high-speed finish cuts and light roughing cuts on most steels and all cast irons. Titanium nitride coatings are very hard and have the strength characteristics to perform well under heavy rough-cutting conditions. All three coatings will perform well on most steels, as well as on cast iron.

9. Name five different insert shapes in order of increasing strength.

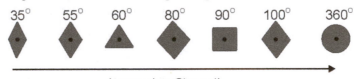

11. State the purpose of tool-nose radius.

    The nose radius of the tool directly affects tool strength and surface finish, as well as cutting speeds and feeds. The larger the nose radius: the stronger the tool. If the tool radius is too small, the sharp point will make the surface finish unacceptable, and the life of the tool will be shortened. Larger nose radii will give a better finish and longer tool life and will allow for higher feed rates. If the tool nose is too large, it can cause chatter. It is usually best to select an insert with a tool-nose radius as large as the machining operation will allow.

13. Describe the three types of side relief angles.

    The side relief angle, also known as the side rake angle, is formed by the top face of the cutting tool and side cutting edge. The angle is measured in the amount of relief under the cutting edge

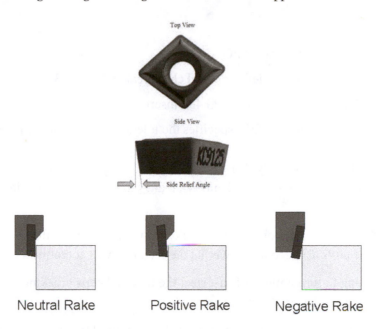

Top View

Side View

Side Relief Angle

Neutral Rake   Positive Rake   Negative Rake

15. What is lead angle?

Lead angle, or side-cutting edge angle, is the angle at which the cutting tool enters the work. The lead angle can be positive, neutral, or negative. The tool holder always dictates the amount of lead angle a tool will have.

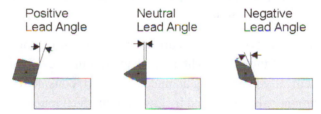

Positive Lead Angle   Neutral Lead Angle   Negative Lead Angle

Lead or side-cutting edge angle is determined by the tool holder type. The lead angle can be positive, neutral, or negative. Tool holders should be selected to provide the greatest amount of lead angle that the job will permit. There are two advantages to using a large lead angle. First, when the tool initially enters the work, it is at the middle of the insert where it is strongest, instead of at the tool tip, which is the weakest point of the tool. Second, the cutting forces are spread over a wider area, reducing the chip thickness.

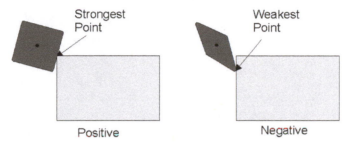

Strongest Point   Weakest Point

Positive   Negative

Increasing the lead angle will reduce tool breakage when roughing or cutting interrupted surfaces.

17. Describe the characteristics of a TPG 432 insert.

Triangular shape, positive relief, Hole and chip breaker, ½" IC, 3/16 thick, 1/32 corner radius.

19. Describe the characteristics of a CNMG 432 insert.

Circular shape, negative relief, M specifies the tolerances, hole and chip breaker, ½" IC, 3/16 thick, 1/32 corner radius.

21. While face milling a previously drilled surface you notice that the inserts start chipping as soon as the face milling cutter enters the interrupted cut. What are some possible remedies for this type of tool failure?

Interrupted cuts can be a problem and should be avoided, use a softer or tougher grade of insert. Increasing the cutting speed and lowering the feed rate will sometimes eliminate chipping.

23. While attempting to turn a piece of cast steel the tool tip keeps breaking off. What are some possible remedies for this type of tool failure?

The major causes of insert breakage are a lack of rigidity, too hard of an insert grade, and low operating conditions. When making roughing cuts through hard spots or sand inclusions, use a tougher, not harder, grade of carbide.

25. While turning tool steel you notice that there is a groove appearing on the insert. What are some possible remedies for this type of tool failure?

The type of tool failure that is occurring is called depth-of-cut notching. Depth of cut notching is an unnatural chipping away of the insert right at the depth of cut line. Depth-of-cut notching is usually a grade selection problem. If you are using an uncoated insert, consider changing to a coated insert. Depth-of-cut notching may also be solved by lowering the feed rate and/or by reducing the lead angle.

Chapter 3 Questions

1. Describe the function of the tool changer.

   Tool changers are an automatic storage and retrieval system for the cutting tools. They can very quickly change tools automatically when commanded to do so.

3. Which axis always lies in the same plane as the spindle?

   Z

5. What piece of equipment gives a machining center a 4th axis of motion?

   A rotary table.

7. Describe the workpiece coordinate or zero point.

   The workpiece coordinate or program zero is the point or position from which all of the programmed coordinates are established. For example, when the programmer looks at the part print and notices that all of the dimensions come from the center of the part, these datums are then used to establish the program zero or workpiece coordinate.

   The workpiece coordinate system can be set from any datum feature. Pick the feature that would allow the programmer to do the least number of calculations.

   The part origin is the X0, Y0, Z0 location of the part in the rectangular or Cartesian coordinate system. In absolute programming, all of the tool movements would be programmed with respect to this point. If all of the dimensions were located from the center of the bored hole, then that point would become the program zero. During part setup, the X and Y zero position of the part has to be located. Using the hand wheels or other manual positioning devices and an edge finder or probe, the setup person locates the point at which the center of the spindle and the part origin are the same. The "home zero" is then entered as a G-code in the appropriate area of the program or in an offset table.

Chapter 4 Questions

1. What type of work is typically held in a vise?

   The plain vise is used for holding work with parallel sides that will fit into a vise.

3. How are large workpieces typically held?

   Bolted down to the table.

5. What is the most common hole-producing tool?

   HSS drill.

7. What is the purpose and advantage of center drills?

   When drilled holes need to be accurately located, it is advisable to center or spot drill the holes prior to drilling. This spotting or centering is achieved with center or spotting drills. These drills are short, stubby, and rigid and do not flex or deflect, as longer drills have a tendency to. The spot drill produces a small start point that is accurately located. When the hole is drilled, the drill point will follow the starting hole that the spot drill made. This method can produce holes that are reasonably accurate in location.

9. What is a reamer?

   A reamer is a cylindrical tool similar in appearance to the drill. Reamers produce holes of exact dimension with a smooth finish. A slightly smaller hole must be drilled in the part before the hole can be reamed.
   The reamer follows the drilled hole, so inaccuracies in location cannot be corrected by reaming. If accurate location is needed as well as accurate size, boring may be necessary. Reamers are a quick way of producing accurately sized holes. Boring can produce holes of any size with a good finish and can locate them very accurately.

11. Describe the two types of tapping that are done on a machining center.

    There are two ways to tap on CNC machines. One way uses a special tapping head. The tapping head is spring loaded, and the lead of the tap provides the primary feed. The secondary or programmed feed need only be approximate because the spring-loaded head allows the tap to float up and down at the lead rate of the thread. The lead of the tap is the distance that the thread travels in one revolution. The lead of a thread and the pitch of the thread are the same for a regular-single lead thread.
    The second type of tapping is called rigid tapping. Rigid tapping does not require special holders. Precise feed and RPM synchronization are needed, however, to insure undamaged threads. The feed of the tap needs to be calculated. The feed for tapping is calculated by dividing 1 (inch) by the number of threads per inch. The result is then multiplied by the revolutions per minute of the spindle. For example: 1/4-20 UNC tap running at 250 RPM. 1/20 * 250 = 0.05 * 250 RPM = 12.5 inches per minute feed rate. Most modern machining centers are equipped with tapping cycles, which feed the tap down to the programmed depth and then automatically reverse the spindle and feed up, unscrewing the tap from the hole.

13. What is a carbide face milling cutter?

Face milling cutters are widely used because of their ability to take large facing cuts. Face mills range in size from 1-1/4 to 6 inches and up. They have a hole for mounting on an arbor and a keyway to receive a driving key. Carbide face mills have carbide inserts that can be indexed or recycled when they become dull.

Chapter 5 Questions

1. What is the most common CNC language in use today?

   Word address programming.

3. Name three functions that miscellaneous codes control.

M00	Program stop	Non-Modal
M01	Optional stop	Non-Modal
M02	End of program	Non-Modal
M03	Spindle start clockwise	Modal
M04	Spindle start counterclockwise	Modal
M05	Spindle stop	Modal
M06	Tool change	Non-Modal
M07	Mist coolant on	Modal
M08	Flood coolant on	Modal
M09	Coolant off	Modal
M30	End of program & reset to the top of program	Non-Modal
M40	Spindle low range	Modal
M41	Spindle high range	Modal
M98	Subprogram call	Modal
M99	End subprogram & return to main program	Modal

5. What does modal mean?

   G-codes or preparatory functions fall into two categories: modal or nonmodal. Nonmodal or "one-shot" G-codes are those command codes that are only active in the block in which they are specified. Modal G-codes are those command codes that will remain active until another code of the same type overrides it. For example, if you had five lines that were all linear feed moves, you would only have to put a G01 in the first line. The other four lines would be controlled by the previous G01 code. The feed rate was modal in the first example. The feed rate does not change unless a different federate is commanded.

7. Write a line of code to move the spindle 10 inches to the right in the X axis at a feed rate of 10 inches per minute.

   G91 G01 X10.0 F10.0

9. Write a line of code to turn the spindle on in a clockwise direction at 800 RPM.

   M03 S800

11. Write a line of code to change to tool 4.

   M06 T4

13. Describe the difference between G92 and G54 workpiece coordinate settings.

The G54-G59 workpiece coordinate is the absolute coordinate position of the part zero. These are not available on all machines. Six are available, and all serve the same function. This allows the programmer to have six different workpiece coordinates established on a machine. This would be very beneficial for repetitive jobs that could be located at the same position on the machine table. For example, a job that is run once each week might use the G59. The G59 would be used to establish the location for that particular job.

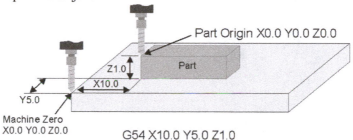

G54 X10.0 Y5.0 Z1.0

The G92 workpiece coordinate is the incremental distance from the workpiece datum (X, Y, and Z zero) to the center of the spindle. In effect, it tells the machine where the spindle is in relation to the workpiece. The spindle must then be in that position when you start to run the program.

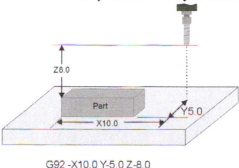

G92 -X10.0 Y-5.0 Z-8.0

G92 workpiece coordinates.

15. Write code to make the following move at a feed rate of 10 inches per minute. Write it in incremental mode. Move 5.000 inches to the right.
G91
G01 X5.000 F10.0

17. Write a line of code using the incremental arc center method to machine the following arc at a feed rate of 5 inches per minute.

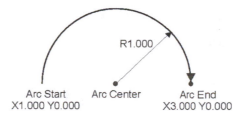

G02 X3.000 Y0.000 I1.0 J0.0 F5.0

19. Write a line of code using the incremental arc center method to machine the following arc at a feed rate of 5 inches per minute.

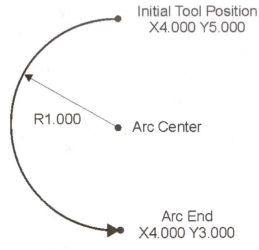

Initial Tool Position
X4.000 Y5.000

R1.000   Arc Center

Arc End
X4.000 Y3.000

G03 X4.000 Y3.000 I0.0 J-1.0 F5.0

21. Write a line of code using the incremental arc center method to machine the following arc at a feed rate of 7 inches per minute.

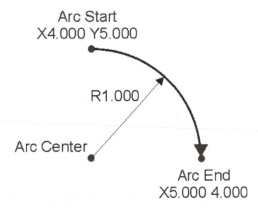

Arc Start
X4.000 Y5.000

R1.000

Arc Center

Arc End
X5.000 4.000

G02 X5.000 Y4.000 I0.0 J-1.0 F7.0

23. Write a line of code using the radius method to machine the following arc at a feed rate of 5 inches per minute.

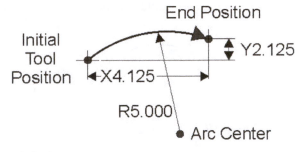

End Position

Initial
Tool
Position

Y2.125

X4.125

R5.000

Arc Center

G02 X4.125 Y 2.125 R5.0 F5.0

278

25. Write a line of code using the radius method to machine the following arc at a feed rate of 5 inches per minute.

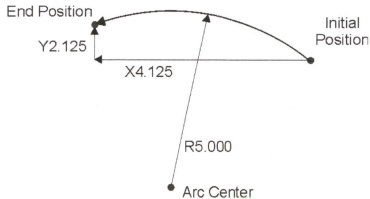

G03 X-4.125 Y 2.125 R5.0 F5.0

27. What must be done to invoke tool diameter compensation?

Cutter compensation can be to the right or to the left of the part profile. To determine which offset you need, imagine yourself walking behind the cutting tool. When compensation to the left is desired, a G41 is used. When compensation to the right is desired, a G42 is used. When using the cutter compensation codes, you need to tell the controller which offset to use from the offset table. The offset identification is a number that is placed after the direction code. A typical cutter compensation line would look like this: G41 D12;.

To initialize cutter compensation, the programmer has to make a move (ramp on). This additional move must occur before cutting begins. This move allows the control to evaluate its present position and make the necessary adjustment from centerline positioning to cutter periphery positioning. This move must be larger than the radius value of the tool. The machine corrects for the offset of the tool during the ramp move. In the figure below the machine compensates for the offset in the move between point 1 and point 2.

To cancel the cutter compensation and return to cutter centerline programming, the programmer must make a linear move (ramp off) to invoke a cutter compensation cancellation (G40). This is an additional move after the cut is complete (point 5 to point 6).

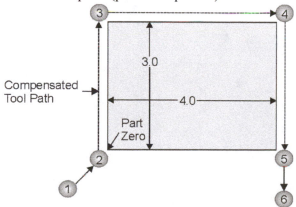

29. Program the part shown in the figure below. Use a .5 end mill to machine the outside shape of the part and machine to a depth of .35. Assume it is tool number 5. Make sure you use offsets. Program using climb milling technique.

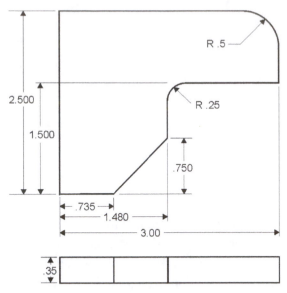

O061
N10 G17 G40 G49 G80 G90
N15 T1 M6
N20 G54
N25 S1500 M3
N30 G0 X-.5 Y-.5
N35 G43 H1 Z1.
N40 Z.1
N45 G1 Z-.35 F10.
N50 G41 D1 X0. F6.
N55 Y2.5
N60 X2.5
N65 G2 X3. Y2. I0. J-.5
N70 G1 Y1.5
N75 X1.75
N80 G3 X1.5 Y1.25 I0. J-.25
N85 G1 Y.75
N90 X.735 Y0.
N95 X-.5
N100 G40 Y-.375
N105 Z.1 F10.
N110 G28
N115 M30
N120
%

Chapter 6 Questions

1. Program the part shown below. Use a .5 end mill to machine the outside shape of the part and machine to a depth of .35. Assume it is tool number 5. Make sure you use height and diameter offset compensation. Program using conventional milling (counter-clockwise).

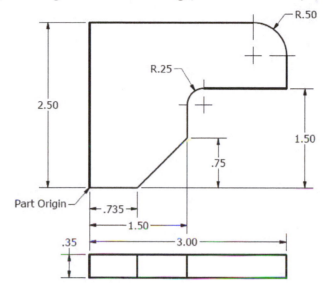

O00061
N10 G17 G40 G49 G80 G90
N15 T5 M06
N20 G54
N25 S1500 M03
N30 G00 X-0.5 Y-0.5
N35 G43 H05 Z1.
N40 Z0.1
N45 G01 Z-0.35 F10.
N50 G42 D05 Y0. F6.
N55 G01 X0.735
N60 G01 X1.5 Y0.75
N65 G01 Y1.25
N70 G02 X1.75 Y1.5 I0.25
N75 G01 X3.
N80 G01 Y2.
N85 G03 X2.5 Y2.5 I-0.5
N90 G01 X0.
N95 G01 Y-0.5
N100 G40 X-0.5
N105 Z0.1 F10.
N110 G28
N115 M30
N120
%

3. Program the part shown below. Use a .5 inch end mill. Mill the contour to a .50 depth.

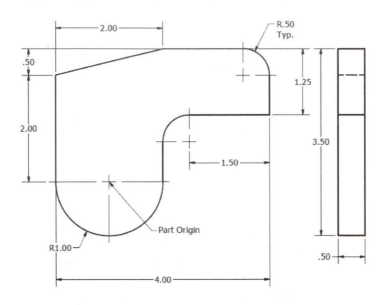

O63
N10 G17 G40 G49 G80 G90
N15 T1 M6
N20 G54
N25 S1000 M3
N30 G0 X-1.5 Y1.
N35 G43 H1 Z1.
N40 Z.1
N45 G1 Z-.5 F10.
N50 G41 D1 X-1.
N55 Y2.
N60 X1. Y2.5
N65 X2.5
N70 G2 X3. Y2. I0. J-.5
N75 G1 Y1.25
N80 X1.5
N85 G3 X1. Y.75 I0. J-.5
N90 G1 Y0.
N95 G2 X-1. Y0. I-1. J0.
N100 G1 Y2.5
N105 G40 X-1.5
N110 Z.1
N115 G28
N120 M30
%

Chapter 7 Questions

1. What type of information does the programmer get from the part drawing?

    The part drawing provides detailed information about the part. The shape of the part, the tolerances, material requirements, surface finishes, and the quantity required all have an impact on the program. From the part drawing, the programmer will determine what type of machine is required, work-holding considerations, and part datum (workpiece zero) location. Once these questions have been answered, the programmer can develop a process plan.

3. What is a process plan?

    Process planning involves deciding when certain machining operations will take place. Primary machining operations will take place on the CNC machine. The part configuration will normally determine the sequence of operations. There are some general rules to follow when deciding on the order of machining operations. The recommended procedures for machining are as follows:
    1. Face mill the top surface
    2. Rough machine the profile of the part
    3. Rough bore
    4. Drill and tap
    5. Finish profile surfaces
    6. Finish bore
    7. Finish reaming

    Note that all roughing operations took place first: then finishing operations were done. This minimizes the effect of high-pressure operations moving or stressing the part. Sometimes secondary operations may be done more economically on other types of machines. Secondary operations often include deburring the part after machining. Secondary operations may be done by the same operator while another part is running. This keeps the CNC machine making parts while the operator does the less demanding secondary operations on a simpler, less expensive machine.

    This approach to process planning is usually done by the operator/programmer in smaller job shop settings. In large shops, the process plan would come down from the engineering area and would include information for each step in the manufacturing of the part. In a small job-shop setting, the operator may act as the manufacturing engineer.

Acme Machining Inc.- Process Plan		Part Datum			Part #
Operation	Tool #	Tool Description		RPM	Feed rate

5. What are the advantages of using a canned cycle?

Canned cycles can dramatically simplify programming and make shorter, more efficient programs.

7. Calculate the feed rate for a 1/2-13 tap running at 600 RPM.

46.15 IPM

9. Write a program to machine the part. Use canned cycles where appropriate.

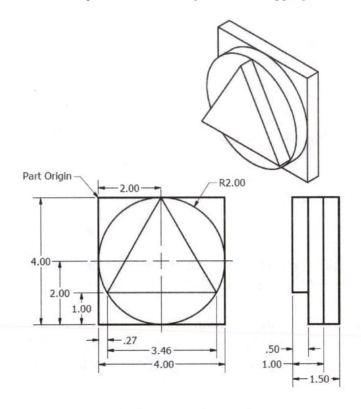

Part Origin

2.00

R2.00

4.00

2.00

1.00

.27

3.46

4.00

.50

1.00

1.50

O79
N10 G17 G40 G49 G80 G90
N15 T1 M6
N20 G54
N25 S1000 M3
N30 G0 X-.75 Y-5.
N35 G43 H1 Z1.
N40 G0 Z.1
N45 G1 Z-1. F10.
N50 G41 D1 X0. Y-5.
N55 X.0 Y-2.
N60 G2 X0. Y-2. I2. J0.
N65 G1 X0. Y.75
N70 G40 X-1.0 Y.75
N75 Z.1

```
N80 G0 Z1.
N85 X-2. Y-4.0
N90 Z.1
N95 G1 Z-.5 F10.
N100 G41 D1 Y-3.0
N105 X.27 Y-3.
N110 X2. Y0.
N115 X3.73 Y-3.
N120 X-.75
N125 G40 Y-4.
N130 Z.1
N135 G0 Z1.
N140 M5
N145 G28
N150 M30
%
```

11. Write a program to machine the part. Use canned cycles where appropriate. Mill the contour .5 deep.

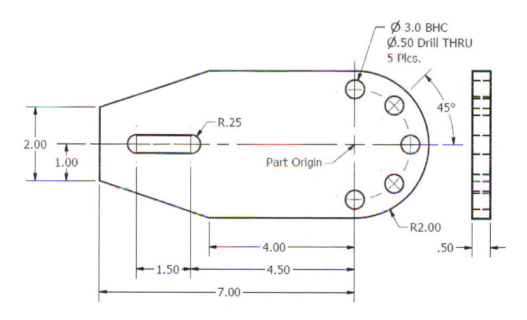

```
O711
N10 G17 G40 G49 G80 G90
N15 T1 M6
N20 G54
N25 S600 M3
N30 G0 X-8. Y-2.5
N35 G43 H1 Z1.
N40 Z.1
N45 G1 Z-.5 F5.
N50 G41 D1 X-7. F8.
N55 Y0.
```

```
N60 Y1.
N65 X-4. Y2.
N70 X0.
N75 G2 X0. Y-2. I0. J-2.
N80 G1 X-4.
N85 X-7. Y-1.
N90 Y2.5
N95 G40 X-8.
N100 Z.1 F10.
N105 G0 Z1.
N110 M5
N115 G28
N120 M01
N125 T2 M6
N130 S1500 M3
N135 G0 X-5.25 Y0.
N140 G43 H2 Z1.
N145 Z.1
N150 G1 Z-.5 F3.
N155 G41 D2 Y.25 F5.
N160 X-6.
N165 G3 X-6. Y-.25 I.0 J-.25
N170 G1 X-4.5
N175 G3 X-4.5 Y.25 I.0 J.25
N180 G1 X-5.25
N185 G40 Y0.
N190 Z.1 F10.
N195 G0 Z1.
N200 M5
N205 G28
N210 M01
N215 T3 M6
N220 S500 M3
N225 G0 X0. Y1.5
N230 G43 H3 Z1.
N235 G98 G81 Z-.65 R.1 F4.
N240 X1.0607 Y1.0607
N245 X1.5 Y0.
N250 X1.0607 Y-1.0607
N255 X0. Y-1.5
N260 G80
N265 M5
N270 G28
N275 M30
 %
```

CHAPTER 8 QUESTIONS

1. Name four of the main components that make up the CNC turning center.

   Headstock, tailstock, tool turret, bed, carriage.

3. State an advantage of a slant-bed-style CNC turning center over the flat-bed-style CNC lathe.

   The bed of most turning centers lie at a slant to accommodate chip removal.

5. What is the most common type of work-holding device used on the turning center?

   Chuck

7. Which machining operation cuts the finished part off the rough stock?

   Parting

9. What is a geometry offset?

   Geometry offsets, or workpiece coordinates, are used to tell the control where the workpiece is located.

11. From what location is the workpiece zero or geometry offset typically calculated?

    The workpiece coordinate is the distance from the tool tip, at the home position, to the workpiece zero point. The workpiece zero point is normally located at the end and center of the workpiece or at the chuck face and center of the machine.

13. How can chips be automatically removed from the bed of the turning center?

    Chip conveyers automatically remove chips from the bed of the machine. The chips produced by machining operations fall onto the conveyer track and are transported to scrap or recycling containers.

15. What tasks do override switches perform?

    Spindle speed and feed rate overrides are used to speed up or slow down the feed and speed of the machine during cutting operations.

Chapter 9 Questions

1. Which turning center axis controls the diameter of the part?

   X

3. What secondary axes addresses are used to define the centerpoint position when cutting radii on a turning center?

   I,K

5. Use the part print to fill in the point locations.

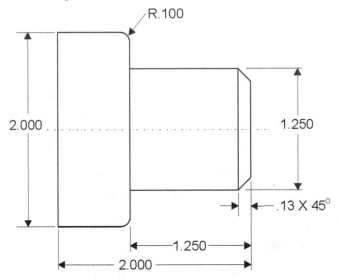

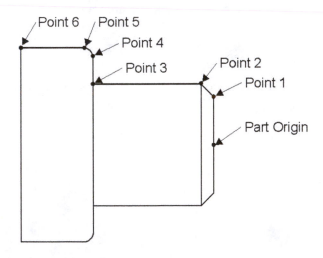

Point #	X (Diameter) Value	Z Value
Point 1	.990	0
Point 2	1.250	.130
Point 3	1.250	-1.250
Point 4	1.800	-1.250
Point 5	2.000	-1.350
Point 6	2.000	-2.000

7. Fill in the blanks to complete the program to machine the profile of the part in question 6.

O0008

N10 G90 G20 G40

N15 G28 U0.0 W0.0

N20 G00 T0101

N25 G50 X10.00 Z12.00

N30 G00 Z.1

N35 G00 X.99

N40 G50 S3600

N45 G96 S200 M03 Spindle Forward

N50 G01 G42 Z0. F.01 Tool Nose Radius Compensation Right, lead in to (POINT 1)

N55 G01 X1.25 Z-.13   X Coordinate at (POINT 2)

N60 G01 Z-1.25 Z Coordinate at (POINT 3)

N65 G01 X1.80 X Coordinate at (POINT 4)

N70 G03 X2.0 Z-1.35 K-.1   CCW arc to X and Z Coordinate at (POINT 5)

N75 G01 Z-2.0 Z Coordinate at (POINT 6)

N80 G01 G40 Cancel Tool Nose Radius Compensation, lead out

N85 G28 U0. W0. M05 Return to the reference point

N90 T0100

N95 M30

%

9. Use the part print to fill in the point locations. Part Origin is located at the right end and center of the part.

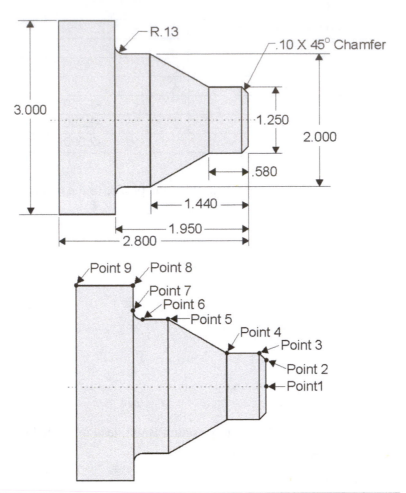

Point #	X (Diameter) Value	Z Value
Point 1	0	0
Point 2	1.05	0.0
Point 3	1.25	-.10
Point 4	1.25	-.58
Point 5	2.00	-1.44
Point 6	2.00	-1.82
Point 7	2.26	-1.95
Point 8	3.00	-1.95
Point 9	3.00	-2.80

CHAPTER 10 QUESTIONS

1. True or **False**? Canned cycles are exactly the same on all CNC machines.

3. If a part needs to be roughed out of bar stock a **G71** would be used.

5. Where must the tool be positioned prior to calling a roughing canned cycle?

   In the first step, the tool needs to be positioned to the rough stock boundaries.

7. A **G82** could be used to spot drill a hole or to counterbore a hole.

9. What is the letter address command to take .100 of an inch off the diameter of the part per pass when using a roughing cycle?

   D.100

11. What letter address controls the pitch or lead of the thread when using a G76 thread-cutting cycle?

   F

13. Program a roughing and finish canned cycle for the part shown in Figure 10–38. Use canned cycles where appropriate and the tooling shown in Figure 10–37. The rough bar stock is 1.30" in diameter.

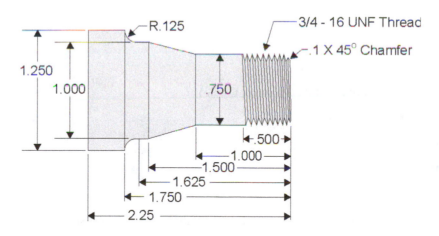

Figure 10–38. Use with question 13.

O1002

N10 G20 G90 G40

N15 G28 U0.0 W0.0

N20 T0505

N25 G96 S200 M03

N30 G50 X8.25 Z12.50

N35 G50 S3600

N40 G0 Z.10

N45 G0 X1.35

N50 G71 P55 Q90 U.02 W.01 D.05 F.01 (Roughing Cycle)

N55 G0 X.550

N60 G1 Z.0

N65 X.75 Z-.10

N70 G1 Z-1.0

N75 G1 X1.0 Z-1.5

N80 G1 Z-1.625

N85 G2 X1.25 Z-1.75 R.125

N90 G1 Z-2.25

N95 G28 U0.0 W0.0

N100 T0303

N105 G96 S300 M03

N110 G50 X8.35 Z12.55

N115 G50 S3600

N120 G0 Z.10

N125 G0 X1.35

N130 G70 P55 Q90 F.008 (Finishing Cycle)

N135 G28 U0.0 W0.0

N140 T0300

N145 M30

%

15. Program a drill canned cycle to drill a 1" hole through the part shown in Figure 10–39. Use the 1" drill from the tool table Figure 10-37.

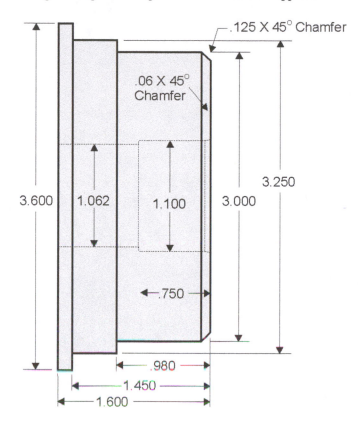

Figure 10–39.

O01003

N10 G90 G20 G40

N15 G28 U0.0 W0.0

N20 G00 T0707

N25 G97 S200 M03

N30 G00 Z.200

N35 G00 X.83

N40 G76 X.675 Z-.50 K.0375 D.0120 F.0625 A60

N45 G28 X2.00 M05

N50 T0500

N55 M30

%

17. Program boring canned cycles to rough and finish bore the 1.062 and the 1.100 bores in the part shown in Figure 10-39. Use G71 and G70 canned cycles. Use the boring bar from the tool table shown in Figure 10-37.

O1005

N5 G90 G20 G40

N10 G28 U0.0 W0.0

N15 G00 T0808

N20 G50 S1500

N25 G96 S400 M03

N30 G00 X1.00 Z.100

N35 G71 P40 Q65 U-.03 W.01 D.030 F.01

N40 G00 X1.22 M8

N45 G01 Z0 F.01

N50 G01 X1.10 Z-.06

N55 G01 X1.10 Z-.750

N60 G01 X1.062 Z-.750

N65 G01 X1.062 Z-1.65

N70 G96 S500 M03

N75 G00 X1.00 Z.100

N80 G70 P40 Q65 F.008

N85 G28 U0.0 W1.0 M05

N90 T0800

N100 M30

%

19. Program a grooving canned cycle to machine the grooves on the part in Figure 10-41. Use the tool from the table in Figure 10-37.

O1007
N10 G90 G20 G40
N20 G28 U0.0 W0.0
N30 G00 T1212
N40 G96 S200 M03

N50 G00 X2.0 Z-1.125

N60 G75 X1.70 Z-1.325 I.125 K.10 F.003 (Grooving cycle)

N70 G28 X2.00 M05

N80 T1200

N90 M30

%

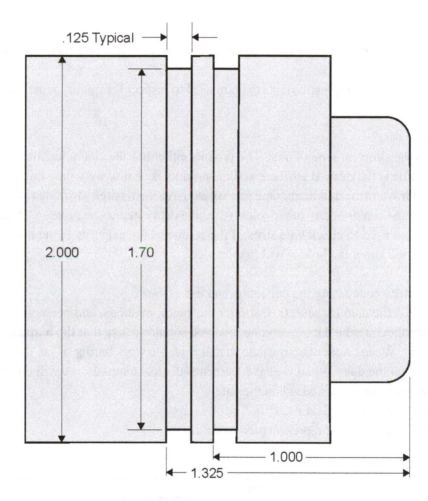

Figure 10-41.

CHAPTER 11 QUESTIONS

1. What is the ISO 9001:2015 standard?

    ISO 9001 is an international standard that specifies requirements for a quality management system (QMS). Organizations can use the ISO standard to demonstrate the ability to consistently provide products and/or services that meet customer and regulatory requirements.

3. True or **false**? ISO 9001 is an American quality management standard.

5. **True** or false? Calibration is done to make sure all measuring instruments are accurate in an enterprise.

7. **True** or false? An operator cannot use a measuring instrument that is past its calibration date.

9. **True** or false? All measuring instruments that are used to inspect for quality must be included in the calibration log.

11. What is attribute data?

    Attribute is the simplest type of data. The product either has the characteristic (attribute) or it doesn't. If blue is the desired attribute and the product is chairs, we would have two piles of products after we inspected them. One pile would have the desired attribute (blue), and the other pile would have chairs of any other color. Attribute data can also be go/no-go type data. Go/no-go gages are often used to check hole sizes. If the go end of the gage fits in the hole and the no-go end doesn't, we know the hole is within tolerance.

13. What are some rules concerning the collection and use of data?

    Make sure that the data is accurate. Consider the gages, methods, and personnel. Clarify the purpose of collecting the data. Everyone involved should realize that the purpose is quality improvement. We are not collecting data to make people work harder or get them in trouble. Take action based on the data. When we have learned statistical methods, we will only make changes or adjustments to processes based on the data.

15. What is a histogram and why is it used?

    A histogram is a graphical representation of data.

17. Code the following:

Blueprint Specification = 1.2755			
1.2752 = -3	1.2749 = -6	1.2752 = -3	1.2754 = -1
1.2759 = 4	1.2750 = -5	1.2761 = 6	1.2752 = -3
1.2748 = -7	1.2756 = 1	1.2752 = -3	1.2756 = 1
1.2753 = -2	1.2755 = 0	1.2747 = -8	1.2749 = -6

19. Code the following:

Blueprint Specification = 5.00				
5.03 = 3	5.06 = 6	5.02 = 2	5.02 = 2	5.07 = 7
5.02 = 2	5.05 = 5	5.01 = 1	5.04 = 4	5.03 = 3
4.98 = -2	4.97 = -3	5.00 = 0	5.01 = 1	4.93 = -7

21. Using the data from question 20, construct a histogram.

```
                              X
                              X  X               X
                    X  X      X  X  X  X  X  X  X
          _____
          -9 -8 -7 -6 -5 -4 -3 -2 -1 0  1  2  3  4  5  6  7  8
```

23. How does an assignable cause of variation differ from a chance cause of variation?

The amount of salt dropped also varied. Could we expect a person to drop exactly 3 ounces of salt every time? The amount of salt is also an assignable cause. How could we remove this cause of variation? Statistical methods can be used to identify assignable causes of variation. Chance causes of variation always exist. Chance causes of variation are those minor reasons which make processes vary. We cannot quantify or even identify all of the chance causes.

25. Consider the following data. Blueprint specification = 2.250.

1	2	3	4	5	6
2.252 = 2	2.249 = -1	2.248 = -2	2.252 = 2	2.246 = -4	2.250 = 0
2.252 = 2	2.249 = -1	2.254 = 4	2.253 = 3	2.248 = -2	2.249 = -1
2.253 = 3	2.243 = -7	2.252 = 2	2.251 = 1	2.247 = -3	2.252 = 2
2.250 = 0	2.248 = -2	2.251 = 1	2.250 = 0	2.246 = -4	2.248 = -2
2.252 = 2	2.249 = -1	2.247 = -3	2.250 = 0	2.249 = -1	2.251 = 1

  a. Code the data. (Hint: 1 should equal 2.251.)

  b. Find the sample standard deviation.   2.48

27. Draw a bell curve and label with standard deviations and mean. Label the percentages for each deviation. Note: You don't have data concerning the actual mean and standard deviation. Draw a generic bell curve.

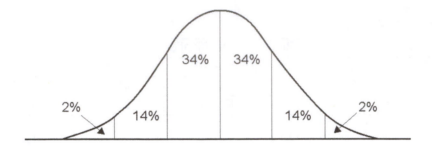

29. Consider the following data.

     9, 0, 1, –1, 2, 8, –1, 0, 0, 3, 7, 4, 2, –3, 8, 2, –4, 3, –5, 5, 3, –2, –2, 0, 3

    a. Calculate the mean. 1.77

    b. Calculate the sample standard deviation.   3.65

31. A machining process is studied and the mean and standard deviation were calculated: Mean = 5, s = 2 (coded data).

    a. Draw a bell curve.

    b. Draw lines where the 6 standard deviations would be.

    c. Label them with actual sizes from this process. (Hint: 99.7 percent of all parts should lie between –1 and +11.)

    d. Label the percentages.

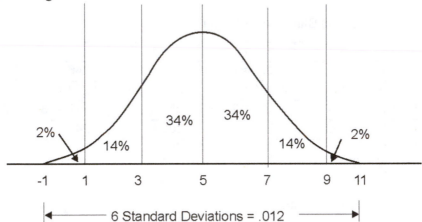

33. Consider the following coded data.

    3, 3, 3, 3, 5, 4, 2, 4, 2, 1, 5, 4, 3, 2, 4, 6, 1, 4, 3, 2

    a. Construct a histogram.

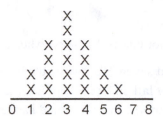

```
            X
            X
        X   X   X
        X   X   X
    X   X   X   X   X
    X   X   X   X   X   X
   _____
    0  1  2  3  4  5  6  7  8
```

b.   Does it look like a normal distribution?

   Yes it does look like a normal distribution.

Chapter 12 Questions

1. List and explain the three rules for charts that show when a process has changed.

   There is a very small chance that seven in a row would be on one side of the centerline. A process can be adjusted if seven in a row fall on one side of a centerline. Assume seven part sizes in a row fall below center. The odds are so low that this could happen that we could assume that something has changed in the process. This is the second rule: if seven fall on one side of the centerline, the process has changed and an adjustment or change is necessary. If the average of the seven sizes was calculated, it would give the exact adjustment needed. The chart not only tells when to adjust, but also how much to adjust.

   In other words, if seven in a row are above or below center, it means that the mean (average) has shifted up or down.

   The only other rule is that a process has changed if seven in a row increase or decrease. This is called a *trend*. If each of the seven in a row gets larger (or smaller), this trend means that the process has changed.

3. These charts make correct decisions **99.7%** percent of the time.

5. A machine has a standard deviation of .003. We are considering running a job on the machine that has a tolerance of ±.006.

   a. Calculate the CP.

   Six standard deviations on this machine would be .018. The tolerance for the job is a total of .012. The capability would only be 66%. The would be a poor job and would result in scrap being run.

   b. Will there be scrap? If so, how much? (Hint: Draw a bell curve to help find the answer.)

   There will be approximately 4% scrap.

7. Standard deviation = .0015; blueprint tolerance = ±.005.

   CP = 1.11 B/C

9. Based on the data from question 8, will this job run well? If you were asked whether to bid on the job, list at least three alternatives you could give.

   No this job will produce at a minimum about 4% scrap.

   They could bid high knowing they would have to inspect and reject parts. They could change the process to make it more capable. They could contact the customer and see if the tolerance can be larger. They could run the par on a more capable machine.

11. Complete an $\overline{X}R$ chart for the process in question 10 (enter the data, calculate control limits, and plot the data).
    See the chart for the answers.
    a. Standard deviation
       2.64
    b. Process mean

c.   Average range

d.   Is the data normally distributed? (Check with a histogram.)

The histogram shows that the process is basically normal. It does show some evidence of a bimodal distribution. It would be wise to run some additional parts and check again.

e.   Upper control limit averages

f.   Lower control limit averages

g.   Upper control limit range

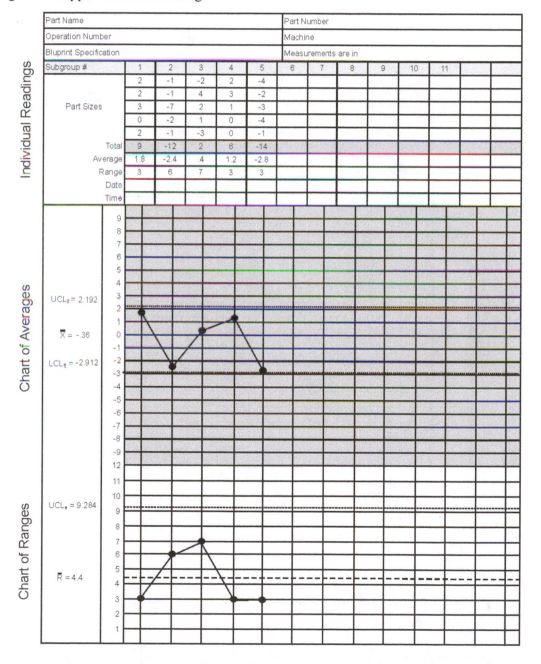

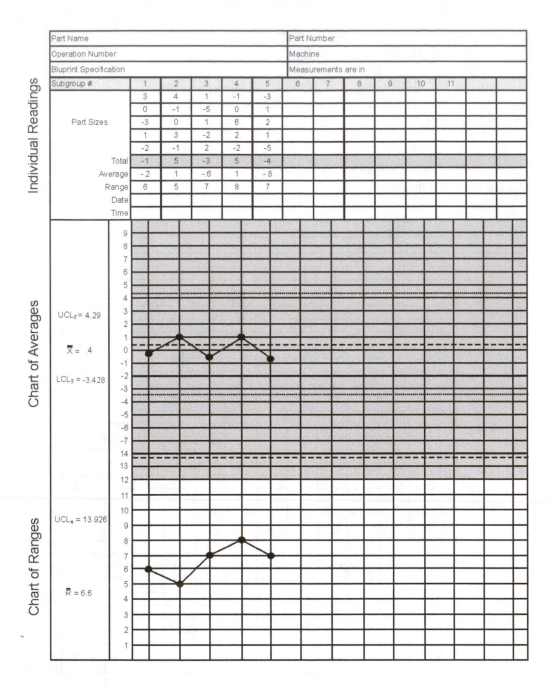

		Part Name					Part Number				

Individual Readings

	Subgroup #	1	2	3	4	5	6	7	8	9	10	11		
		3	4	1	-1	-3								
		0	-1	-5	0	1								
Part Sizes		-3	0	1	6	2								
		1	3	-2	2	1								
		-2	-1	2	-2	-5								
	Total	-1	5	-3	5	-4								
	Average	-.2	1	-.6	1	-.8								
	Range	6	5	7	8	7								
	Date													
	Time													

Part Name — Part Number

Operation Number — Machine

Bluprint Specification — Measurements are in

Chart of Averages

$UCL_{\bar{x}} = 4.29$

$\bar{\bar{X}} = .4$

$LCL_{\bar{x}} = -3.428$

Chart of Ranges

$UCL_R = 13.926$

$\bar{R} = 6.6$